together, closer

PENGUIN BOOKS

together, closer

GIOVANNI FRAZZETTO was born and grew up in the southeast of Sicily. He studied science at University College London and received a PhD from the European Molecular Biology Laboratory in Heidelberg, Germany. The author of *Joy, Guilt, Anger, Love*, Frazzetto lives in Ireland.

ALSO BY GIOVANNI FRAZZETTO

Joy, Guilt, Anger, Love

together, closer

The ART and SCIENCE of INTIMACY in FRIENDSHIP, LOVE, and FAMILY

GIOVANNI FRAZZETTO

PENGUIN BOOKS

PENGUIN BOOKS

An imprint of Penguin Random House LLC
375 Hudson Street
New York, New York 10014
penguin.com

LIBRARY OF CONGRESS CATALOGING-IN-PUBLICATION DATA

Names: Frazzetto, Giovanni, author.
Title: Together, closer : the art and science of intimacy in friendship,
love, and family / Giovanni Frazzetto.
Description: New York : Penguin Books, [2017] | Includes
bibliographical references and index.
Identifiers: LCCN 2017003383 (print) | LCCN 2017019494 (ebook) |
ISBN 9781101992227 (ebook) | ISBN 9780143109440 (paperback)
Subjects: LCSH: Intimacy (Psychology) | Interpersonal relations.
Classification: LCC BF575.I5 (ebook) | LCC BF575.I5 F73 2017 (print) |
DDC 158.2—dc23
LC record available at https://lccn.loc.gov/2017003383

Printed in the United States of America

1 3 5 7 9 10 8 6 4 2

Set in Goudy Oldstyle   •   Designed by Elke Sigal

For my nieces, Alice and Eva

CONTENTS

together,
closer

Prologue

This book is about intimacy and why we need it. Mixing narration and science, it tells stories about relationships.

As human beings, we have a penchant to connect. Like waves cling to the shore, so we are inclined to attach. There may be seasons of low tide, an occasional desire to drift solo, or storms that strand us, but eventually we will seek or return to a harbor. Loneliness can kill, whereas togetherness revives. We live in a world where it is much easier to find isolation than companionship. Yet meaningful relationships are the most nurturing ingredient for our happiness.

Intimacy eludes singular definitions. From casual sex to lifelong bonds, from marriage to betrayal, from friendships to unconditional love, when we witness birth or death, intimacy reclothes itself constantly.

Much as we crave intimacy, we may also be in awe of it. It is an unmasking form of mutual knowledge and belonging, which many of us may go to great lengths to shun.

If refracted through the prism of science, the experience of intimacy emerges in the most mundane fragments of our daily lives. It is about how we perceive with all our senses; how we steer our minds and carry our bodies in relationship to others; how we seek and offer rewards; how we predict risks and make decisions; how we fear or encourage; how we create memories; how we trust and make ourselves vulnerable; how we learn. While intimacy is a plausible object of scientific inquiry, it is also an important thread of lived experience.

In this book we will encounter characters whose fears and desires will usher them in, through, and out of intimacy. A single woman in her forties at odds with the uncertainty of finding a partner. A husband who looks back in time to the beginning of his marriage. A man and a woman entangled in a secret love affair. Incompatible partners caught in a chase of union and separation. City daters striving to unite love and sex. A father and a daughter whose relationship shines as his death nears. Two men who figure out what they can teach each other. Inseparable friends who together sketch the path of their future intimate lives. Where possible, their thoughts, feelings, and actions are explained through notions and experiments in biology, psychology, and neuroscience. Life wisdom is mixed with knowledge of the body and the mind.

Through these stories, we are invited to reflect on our own experience of intimacy. How we reach it and lose it. How we see it disappear or grow as we also transform, and renovate the way we love. How we get close, and closer.

Shidduch

S now muffled the start of a new day. The river was frozen, and the view from Anita's window was reliably gray.

When her eyes opened, she found herself snug like a baby, as if tightly curled around the spool of some dream she could neither remember nor let go, her head buried in her breast, her arms around her knees. She took two or three deep breaths to broaden herself and was grateful she had not woken up coughing during the night—the new herbs from Inge, whom she called her infallible witch, were doing their job.

Slowly she unwound and stared at the ceiling. She wondered if she had closed the bath tap completely before going to bed, and reached her hand to the floor to check for any flooding. Reassured, she gently yawned, sat up, and shook her head.

She closed her eyes again.

"Many sweet returns, Anita, and happy fucking Valentine's Day," she said, acknowledging the cruel game of fate that, ever since she was an adolescent, had made her wish her birthday could be forever scraped off the calendar. Then she grinned at her fat, sleepy, ginger cat.

"Where's my breakfast in bed, Joshua?"

Joshua wasn't the cat. The cat was called Whiskey. Joshua was a fantasy boyfriend she had created.

"Tell me. Is he German?" had been Ruth's first question.

"No, Mother. He's American."

"And is he one of us?"

"Yes, Mother."

"Oh, *motek*, excellent! Is this a serious thing? When are we meeting him? What's his name? And please don't tell me he's also an artist!"

Anita was a photographer. She took pictures of abandoned spaces. Born and raised in Brooklyn, she moved, after art school in the Midwest, to Germany impulsively, when nothing was working for her at home. Berlin, the city whose mayor had once branded "poor, but sexy," was the best haven for every jobless artist on the planet. Now her professional life was enviable, at least in comparison to what she had left behind in the United States. Her work was always being shown on at least two continents. She had a big solo exhibition on the horizon, a part-time teaching appointment at an English-language college, a personal assistant, and regular sales.

She lived in a lofty flat, big enough to use as her studio. She was regularly invited to parties, and she had time to visit galleries and museums for inspiration. She had money to travel, to eat out on weekends, and to buy herself a new handbag from time to time. All considered, there was nothing she needed to worry about.

But even if she was able to see that for herself, it didn't count a dime when in the evenings all she had to cling to was the warmth of a pillow and Whiskey's fur.

She had met a lot of people in her new city, in particular the crowd of expats from the art scene, and that was a good thing for her work. But she couldn't say she had someone she could really count on and that awareness, like a side effect of a new diet regimen, made her uneasy.

Winters were especially harsh and solitary in the German capital: the large avenues muddy and silent, the bakers rude, and most people grumpy and irascible. With luck, commuters on the U-Bahn might smile again, maybe, in mid-April.

Anita created Joshua on a day when three times in a row a stranger in the street didn't return her greeting, a boy didn't show up for a date, and Ruth asked her when she would finally get married. The invention of an imaginary companion helped her deal with her parents' insistent concerns about her still being single but was also a way to mitigate her solitude and fantasize about what she wanted in a man. Occasionally, she would use the lie with strangers, such as when she was the only single person at a dinner party. Like Whiskey, Joshua was a redhead. He had hazel eyes. He was tall and a sculptor, which meant he had large, mighty hands. He was calm and assertive, funny and a bit clumsy. Above all, Joshua was a partner in crime and the comfort Anita missed through the ups and downs of her daily life. She needed someone to tell her off if she spent too much time in front of a computer screen, someone to sit her down at a table each time she ate while standing, a guy who would listen to her when she headed down a slope of negativity or ranted about a client who backed off a sale. She needed someone to remind her that fidgeting made her unattractive and that house fires are possible but rare. A guy who could laugh sweetly and take pictures of her while she danced naked to 1980s music videos and who would make a mess in the kitchen every Saturday, for pancakes, or to bake her a birthday cake. A man she could improvise with and on a whim book tickets for Africa or Brazil just because it felt right in the moment. Most important, Anita needed a man who, as she put it, could "fucking tell her that everything was going to be all right." This was an odd turn of perspective. For years, since she was sixteen or so, she really had not wanted anybody—parents, friends, or boyfriends—to do that. Only she knew what was best for her. Now, in her early forties, she was fed up with being her only point of reference. Even the easiest task took her ages to accomplish. Every decision she made on her own felt like a matter of life and death. Even if he was unreal, she thought, Joshua halved the weight of her worries.

* * *

Before sprawling on a rug, Anita reached for her notebook and a pen. "LONELINESS KILLS," she wrote in block letters, promptly crushing her last pack of cigarettes.

It would make a great T-shirt, she thought. She could use it to come out as lonely at the next party or on the street, in the hopes to start a club or something, and maybe attract like-hearted people.

Loneliness is a world epidemic. A comparative survey has shown that in less than two decades, from 1985 to 2004, at least in the United States, the number of people who had no confidant to talk to about important matters almost tripled. On the other side of the pond, the scenario is not brighter, the UK being among the most lonely countries in Europe.

Anita's slogan is no joke. Loneliness can be a cause of early death, just like smoking, obesity, lack of exercise, or air pollution. It harms our bodies and changes the way we perceive and interact with the world. It induces fatigue and encourages sleep disruption. It is associated with stress, anxiety, and depression. It is connected to elevated blood pressure and to damage to the cardiovascular system. It ignites cellular inflammation and impairs immune defenses. It may also lead to cognitive decline and ultimately to dementia.

Anita felt the burden of loneliness as a relentless grip in her chest. A chronic sufferer of heartburn, she also had dyspnea, or rather what is called pseudo-dyspnea, which meant she hyperventilated, coughed, and was occasionally short of breath.

We steer our minds and carry our bodies in relationship to events and others. We inhabit, share, perceive, and interact with the world through the totality of our bodies, not just with the brain. The vicissitudes of life, and how we react to them, take a toll on our body-mind equilibrium, affecting the function of organs, tissues, cells.

Inge warned Anita: "We need to take care of your parasympathetic nervous system!"

While the sympathetic nervous system is "on" when we need to deal with danger or an emergency, the parasympathetic nervous system dominates when we can afford to relax. It carries out functions that don't need our attention, such as cardiac movements, respiration, and digestion.

If, out of worry or stress, we are constantly vigilant, the parasympathetic nervous system resents it. We can't relax. Not only that, some of the simple automatic functions it carries out go awry too. One of these is the draining of acids from the stomach.

Anita's chest pain and shortness of breath were not only a result of what she ate or how much coffee or gin she gulped. They were also a response to her loneliness, as well as to her preoccupation with it. One crucial component in the parasympathetic nervous system is the vagus nerve, a long nerve that travels from the base of the skull through chest and down to our genitals and is sensitive to social interactions. Among its many functions, the vagus nerve contributes to the regulation of our gastrointestinal tract. If something is awry with the vagus nerve, stomach acids are not drained properly. When acids accumulate and they back up on the esophagus, they burn tissues. The presence of acid on the local nerve endings causes them to signal incorrectly that there is not enough oxygen. As a consequence,

we hyperventilate. Then we feel dizzy and edgy, and the agitation ex-
acerbates our preoccupations and distress. It is a vicious cycle, and
Anita was entangled in it. Inge massaged and acupunctured Anita.
She gave her herbs and commonplace advice to move and take exercise. It was inertia, not exhaustion, that cut Anita's breath short.
Anita did take up jogging, remembering a quote from Erma Bombeck
that went: "The only reason I would take up jogging is so that I could
hear heavy breathing again."

Still in bed on that Valentine's Day morning, Anita suddenly noticed something out of the corner of her eye. It was a man deftly
walking on the gable roof of the house across the street, and for an
instant she believed it was a cat. She turned her mobile on and it
beeped. An e-mail from her mother:

> Surprise!! We are at the airport. Plane lands in Berlin
> around 1 pm, on way to Prague.

> Happy birthday, my child. Can't wait to see you—AND
> Joshua!

> Mami and Baba

Crap! Anita was in trouble. She had to find some perfect excuse
or reveal the big secret.

*   *   *

Before Joshua was invented, Anita's romantic life had been a trou-
bling concern for Ruth. Her sense of accomplishment as a mother de-
pended on it. With apprehension, Ruth thought of her daughter, her
youngest—her "most golden," she used to say—alone in a foreign

country, away from her family, and wondered naively how different it might have turned out if Anita had never left home.

"If Anita were still home, she would definitely be partnered by now."

Then she would produce a list of eligible men, which she had amassed by asking colleagues, relatives, neighbors, and friends at the synagogue. The pressure on Anita to find a partner and settle down was enormous.

Ruth met her husband, Steve, at a Rosh Hashanah dinner, though he had already been eyeing her in the streets of Brooklyn. Marrying out of the Jewish faith was not an option. Inexperienced, just out of high school, when her wedding day arrived, Ruth had her whole family behind her, and she married into another big family. Her marriage wasn't an arranged one, a *shidduch*, but it did conform to standards of parental approval.

Anita dreaded generational comparisons. Family gatherings were invariably an occasion for a horde of older relatives, including her parents, to recite the same old poisonous adage: "How can someone like you be single?"

Beautiful, clever, educated, well-read, with a job and, they didn't refrain from pointing out, from a good family—whatever might push men away from their precious jewel, their little gorgeous Anita?

How annoying. How insensitive of the whole lot, she thought, to rub salt into a young woman's wounds. Not only did they make her condition of being single sound like a failure, but oblivious to the contemporary etiquette of the dating world, they also showed complete ignorance of what it takes to find a partner today. Besides, contemporary urban life in Berlin, or New York, is not what it was in Brooklyn when they were young.

"At your age, not only was I married—" Ruth would begin to say, and without letting her finish her sentence, Anita would continue, "I already had a house, a dog, you, and your brother!"

Anita loathed being alone but knew that being single was common in her circles. Though she didn't disdain it, marriage for Anita was a distant, improbable milestone. For her family, Anita was simply late. The demographic figures across their generational gap are telling: The 1950s, when Ruth and Steve married, were definitely a golden age for nuptials. The median age at first marriage was even lower then than it had been in 1890! On average, women married by the age of 20.5 and men by the age of 24.0. In 2010 that same median age rose to a record high for the past hundred years. It was almost 27 for women and almost 29 for men. The proportion of men and women who by age 35 had never been married was also high, being 14 and 11 percent, respectively.

When Anita mentioned things like sex appeal, or the words "fancy" and "spicy," and how these belonged to her standards of choice when looking at men, Ruth would say to her:

"Do you think I didn't find Baba attractive? I chose the nicest man among those who wanted me and were available. Maybe it was a big illusion, but look where we are now. We saw no option other than sticking together. I didn't know if I was ready—I just let myself go. I suppose it was destiny. Maybe I was lucky."

*   *   *

Luck. It's easy to regard romantic outcomes as if set by an erratic fate.

At the same time, we wish we could corrupt destiny, mold it into a recognizable and convenient pattern that fits our aspirations. We inherently try to silence the incessant scream of uncertainty. We would like to predict beginnings and endings, chain the future, quench our thirst for definite answers even in matters of the heart—as if it were enough to follow a strategy to obtain a desired outcome.

At the park, at parties or in cafés, even at the supermarket, Anita found herself staring at other couples as they stroked each other with blades of grass, made out with their eyes closed, and as they checked off their shopping list.

What did they have that she didn't? What made them ready to connect?

The simple fact that they were in a relationship, and she was single, made them superior in her eyes. She looked up to them as if they were holders of some deep, secret meaning—often forgetting the many couples she knew who were drowning in arguments and incomprehension. In search of clues about her own romantic failures, she would watch them and wonder what traits and attributes might be the successful ingredients of a relationship. Was it tenderness, or was it some invisible complicity? Was it kindness? Was it directness, respect, independence, or the sex from the night before? Typically, after such voyeuristic sessions, she would conclude: "I just don't know how to love."

But then she would hasten to revise her initial judgment with its exact counterpart: "This is ridiculous. Of course I know how to love."

She abandoned herself to the motion of an existential pendulum: Her sense of self-worth was the pivot, swinging from one extreme view of herself to another, from hope to resignation.

She also wondered and worried about something else: How much time she would gain if she had a partner? She believed that people in a relationship had one thing less to worry about. They could better concentrate on other aspects of their lives, not having the problem of being single like a sword above their heads. She was envious of that and afraid her loneliness might interfere with her work. Could other people read on her face that she was lonely? Was that why sometimes clients didn't buy her pictures or she didn't get a job? Might her loneliness have scared people away?

Loneliness obfuscates. It becomes a deceiving filter through which we see ourselves, others, and the world. It makes us more vulnerable to rejection, and it heightens our general level of vigilance and insecurity in social situations.

Too much isolation interferes with our capacity to scan, understand, and interpret emotions. When faced with images depicting four basic emotions—happiness, fear, anger, and sadness—lonely people are less good than nonlonely people at interpreting them. The lonelier they are, the worse their ability to distinguish them. When lonely, we are also less capable of capitalizing on positive experience. Instead of concentrating on the joyful or positive aspects of things, we focus on the negative. We stress more easily. We are less optimistic.

A brain-imaging study showed revealing differences in the ways lonely and nonlonely people reacted to pictures of objects or other human beings, with or without a social theme. When shown pleasing social pictures of people, such as a man running with a dog, the nonlonely participants reacted with greater activation in the brain's reward center. The same didn't happen for lonely people: activity in their reward center was higher in response to pleasing pictures of objects, such as money or a rocket liftoff, and not people, indicating a certain inability to react to, and relish, social stimulation. The results reversed when the pictures used in the study depicted people in danger, such as a soldier or a woman being slapped. In comparison to the nonlonely, the lonely people paid more attention to these pictures, as revealed by higher activation in the visual cortex, but they also showed less empathy or concern, as shown by weaker activity in an area of the brain that helps us intuit other people's mental states or take their viewpoint. As the authors of the study suggest, being less open to pleasure and more acutely vigilant about threats can actually make us lonelier. Loneliness begets loneliness, in a cycle. One of the most pervasive and perverse effects of isolation is that the more time we spend in it, the harder it becomes to overcome it.

Journal writing, years of psychoanalysis, yoga, feminist texts, poetry, acupuncture, chat forums, women's magazines, the Grinberg Method, and Inge's herbs—there wasn't anything Anita hadn't tried to become ready to love. She read the horoscope for Aquarius every day, and she once even resorted to a psychic, who told her love would knock on her door at the age of forty-four, from abroad, in the form of marriage and with the prospect of a child. No divorce.

Without hesitation, Anita sent the prediction to Ruth.

*　　*　　*

In a fascinating analysis, the sociologist Eva Illouz uses the forces at work in market economies to make sense of the impressive shift in the dynamics of choosing and finding a companion today. In a nutshell, she explains that a whole new "ecology" of romantic choice and partner selection has spread in the arena of dating and matchmaking. Rupture with old-fashioned family, community, and religious networks; a craving to find in people a harmonious union of emotional and sexual attractiveness; and the expansion of options to choose from, due for instance to the introduction of Internet dating and the greater availability of casual sex, have changed the way we seek and gauge the eligibility of a potential partner. The larger the number of available options, the slimmer the chances of settling for one or the other. Abundance of choice dilutes appeal because when there are too many possibilities to choose from, it becomes more difficult to appreciate their value. On the other hand, when fewer options are available, their singular appeal becomes more pronounced. So does our desire.

Such an embarrassment of riches was showcased in a study conducted in a food store. Researchers exhibited a selection of high-quality jams with samples available for tasting. In one scenario, customers had six kinds to choose from. In another, twenty-four jams were available. In the end, the stand with the wider selection enticed more customers, but in both cases people tasted about the same number of jams. When encouraged to actually purchase a jar of their choice with the gift of a one-dollar coupon, the reactions were strikingly different. While 30 percent of the people exposed to the smaller selection of jam eventually bought a jar, only 3 percent of those who had to choose between twenty-four different kinds made a purchase.

As in the case of jam jars, an overabundance of partners to choose from prevents commitment.

The French mathematician and writer Blaise Pascal once wrote: "Clarity of mind is clarity of passion, too; this is why a great and clear mind loves ardently and sees distinctly what it loves." It is an attractive equation. How convenient would it be if by simply assessing pros and cons, strengths and weaknesses, we could rationally steer inclination, transform chaos into order? However, rationality, or excessive calculation, can do more harm than good in certain realms of life.

We cannot predict our feelings based on analysis. Equally, we cannot forecast romantic outcomes. Some factors in a decision will always remain hidden from conscious reasoning. In other words, reasoning may stifle emotion and thus not expose our truest intentions, resulting in poor choices.

"I didn't know if I was ready," said Ruth. She just went for it. The societal circumstances around her encounter with her husband, Steve, were different, but her marriage lasted. Psychological research shows that when the course of a relationship is monitored, initial levels of satisfaction, whether promising or unhopeful, match the outcome over time when they are not overanalyzed. Two groups of couples were asked to evaluate their relationships. One group spontaneously offered their impulsive feelings. The other group rated their satisfaction after listing the reasons why they thought their relationships were going well or badly. Months later, the failure or success of the relationship coincided with the initial rating much more for those who expressed their impulsive feeling about it than for those who had brooded about it.

Anita gave the impression of being available, and that no one was interested in her as a partner. But she was also difficult when it came to choosing. Suitors who came her way, or those Ruth suggested to her, were never the right ones. She was at once a victim of and an accomplice in the choice overload. Anita's girlfriends who were already married advised her not to be so picky when it came to men. They spoke as if her situation was so desperate that she should go off with the first bloke she encountered on the street. Interestingly, a study showed that, at least with regard to physical attributes such as weight, height, and body mass, more often than not there are huge discrepancies between what we ideally fantasize our partners to be and who we end up with in reality. Such discrepancies are more accentuated among women, who have been shown to regard physical attractiveness as a less crucial factor in their choices and may place more emphasis on socioeconomic status, for instance.

There are a lot of people who complain about being alone, but they either do nothing about it, or they keep avoiding the reality of finding a partner because of the compromises involved. However, in the end, those who succumb to the embarrassment of choices—and choose no one—feel the most miserable. On the contrary, those who settle for perhaps less than what they expected are substantially happier and more satisfied. Even if settling for someone might be scary because it rules out other opportunities, it is also what ultimately makes us healthy and content.

*   *   *

Anita's intolerance for her parents' harping on the importance of finding a partner arose partly from her awareness that they were right. She just had a hard time admitting it. The comfort we experience when we don't feel alone is irreplaceable.

Authentic, meaningful bonds with people who can be trusted, as opposed to superficial acquaintances or virtual connections, are great providers of health (better than possessions and comfortable financial conditions).

The need for proximity and sensitivity to abandonment are universal across the animal kingdom. Fruit flies live longer if they experience social interaction. Separate a litter of newly born mice or rats from their mother and they will protest loudly, with incessant squeaks and cries, under considerable stress, until they are reunited with her. If the separation is protracted, it will trigger a cascade of physiological consequences that will impair growth and interfere with the behavioral development of the pups.

In this context, touch has a vital role. Deprived of contact, we become hungry for touch. As animals develop—humans included— touch counts even more than nourishment. In the 1950s, Harry Harlow, an American scientist who worked in Madison, Wisconsin, showed this in a set of groundbreaking experiments. He separated infant monkeys from their mothers at birth, then confronted them with two unusual kinds of surrogate mothers. One was a bare structure of wire mesh. The other one, also in wire, was covered with a soft cloth. Although both kinds of mothers could dispense milk through a bottle, the infant monkeys kept clinging to and cuddling with those wearing the cloth. In a subsequent set of experiments, Harlow checked what would happen if the infant monkeys were exposed to stress. In the presence of mechanical teddy bears playing drums, which were employed to scare them, the monkeys rushed to hug and rub against the cloth mothers, as a way to cope with the stress and calm down, regardless which mother they got milk from. In humans, touch represents a form of comfort and a benefit to health throughout life. Newly born infants who were regularly massaged during the day grew almost 50 percent faster and showed behavioral advantages in comparison with infants who were not touched as regularly, despite eating the same amount of food. Hugging lowers blood pressure and boosts our immune system. A study showed that holding their husbands' hands reduced women's discomfort in response to the threat of a mild electric shock. Older adults prospered in good health (for instance, they needed fewer medical appointments) if massaged regularly, and did even better when they were given the opportunity to massage infants.

We live in a society that is touch deprived, but we all need physical contact. It became a recurrent joke among Anita and her other single friends that when someone would finally show up and touch them again in tender, loving fashion, they first would have to bathe with a limescale remover, so thick was the coating of their isolation. From New York to Tokyo, shops have opened where it is possible to buy cuddles. Initiatives like this are symptomatic of an urgent need in our society for connection, the lack of which, as we have seen, has lasting detrimental consequences. Connection needs to be trained.

Recently, a study has identified in the brains of mice a population of neurons with a function that confirms the crucial role of social interaction. When the mice were exposed to an acute period of isolation, these neurons reinforced their synaptic connections, as if together acknowledging or protesting against the new difficult conditions. When the isolated mice were reunited with other mice, the activity of those same neurons peaked high in response to the reestablishment of social contact, and the animals were more sociable in comparison with those who had not suffered from confinement. Interestingly, when the mice who had been kept in isolation were given the chance to stimulate those neurons, they would only do so if they knew they would be in the presence of another mouse, suggesting that when alone the activation of those neurons stimulates an unpleasant sensation of loneliness that they wanted to avoid. Those neurons were a sensitive target for the experience of loneliness, but also a useful resource to recover from it. These subtle, remarkable findings confirm the idea that from shifts in behavioral responses to alterations in brain activity, and from the well-being of single nerves, such as the vagus nerve, to the responsiveness of a distinct population of neurons, the integration of body and mind within the social fabric is a carefully regulated affair that is also highly vulnerable to disruption.

It seems that today we live in a society that is collectively making human interactions extremely difficult to achieve, even though we know they work like magic to improve the quality of our lives. It's indeed a frustrating reality of our species that our most natural inclination to connect should be numbed.

Regardless of whether it turned out to have viable prospects, even the shortest relationship, and the validation she would get from it, would make all Anita's concerns and symptoms disappear. In the absence of that kind of physical and emotional connection, her symptoms would resurface powerfully.

* * *

What to do about Joshua?

Ruth and Steve were soon to show up at Anita's doorstep, and Anita didn't have a lot of time to decide. Was she to bury her fake boyfriend and admit her solitude to her parents or keep him alive and continue to lie? She was torn and felt a little ashamed about the whole story. On one hand, she would have liked to tell her parents it was all an invention, in the hopes to gain understanding and comfort and not their reprimand. On the other, it might have been easier to continue the charade. But how much longer could she have done that? In either case, her solitude would remain. The friend who had occasionally posed with her in selfies, pretending to be Joshua, was out of town, and that was a blessing, she thought, because who knows if he would have been capable of keeping up the farce. She considered the option of saying she and Joshua had just broken up. But that would have meant having to explain why and how or when exactly, as well as deciding whether she was sad or glad about losing him. Too many explanations, bigger drama, even bigger lies.

When Ruth and Steve saw Anita arrive at the restaurant on her own, they didn't have time to say anything:

"Joshua is out of town for work. He won't be able to join us."

"Really? On your birthday?" asked Ruth.

"Yes, an exhibition."

"Well, can't we call him? I would like to talk to him."

Discreetly, Steve poked Ruth's leg under the table and said: "Anita, sweetheart, you'll introduce us to Joshua whenever you are happy to do so. Let's just have a good time here, now. We have missed you."

On their way back to the hotel, Ruth said to Steve, "I have the impression we will never meet Joshua, at least not this one—don't you think?"

"Sure we won't. But let's make her think we didn't realize, until the real one shows up."

"Oh my golden little girl," said Ruth.

*   *   *

Back in her flat, Anita climbed into bed and stared out the window. On the horizon, faint and out of focus, a new year appeared before her.

Investing in an illusory boyfriend had definitely been a waste of energy. It had also been a huge distraction from reality. Anita's biggest problem was her stubbornness in trying to predict and frame her romantic outcomes. She did not yet understand how to cope with uncertainty. She needed to let go of the illusion that she could find the perfect partner or make one exist and, instead, work toward being ready for one. Doing so would probably not make him more likely to show up, but it would make her less vulnerable to disappointment.

When we are constantly deprived of something we are thirsty for, we tend to lose confidence completely. We become anxious about the possibility that our wishes might never be fulfilled. When we feel anxious, it's more difficult to remain hopeful. But when dealing with something entirely out of our control, the best thing to do is to remain open, with all our senses, and to sideline our intentions. This doesn't mean we must suddenly believe in fate or take the first option available. It means we should make room for the unexpected, which is usually around the corner.

By doing so, we also have a chance to refocus our attention on the present and what makes us stronger and happier on our own. Priorities will align better too.

On some days, Anita always had her art as a consolation. On others, the replacement for Joshua would be found in friends, travel, and other daily pleasures. Whiskey wasn't going anywhere.

Perhaps intimacy always begins and ends with ourselves.

The Leap

I t's late afternoon on New Year's Eve, and Aidan is out buying violets for his wife, Carrie. It's their thirty-fifth anniversary, or the twenty-first, depending on how they counted. Going by the latter numeration meant celebrating their bond only every time a leap second was added to time, because it was on the last night of December in 1973 that a leap second tipped their relationship forever.

Since then, life has been generous to them, and to look back at how it all started makes Aidan feel he's a very lucky man. He's picking up the flowers from the same stand where he got Carrie the first rose, roaming the same streets he used to as a lad. Only he walks more slowly now than he did back then, and he is not in a rush. The pub where they used to eat their Sunday roast is still there, and so is the newsagent where Carrie used to buy her lottery tickets once a month. Their local cinema is long gone.

Aidan's mind brims with memories, and his love for Carrie, after all these years, is stronger than ever.

Their eyes locked for the first time thanks to a detour. Aidan was an accountant in Covent Garden. One evening on his walk back from the office, his route home was affected by roadworks. "Bugger," he said. "I hope they'll sort this quickly." That same morning a doctor had delved down his throat with a clipper to extract microscopic morsels of his esophagus.

"Will I ever be able to eat again after this procedure, madam?"

"It might just radically alter the way you crave," the doctor replied.

Because of the diversion, Aidan would have had to walk around a block, cross a busy street, and pass a car park, but it occurred to him that entering the old bookstore through its main door and leaving through its back exit would be a convenient shortcut, saving him a few minutes. Aidan had never been a keen reader. He might have entered the shop once or twice before, only to seek shelter when it rained and spend a few minutes looking at posters of maps or browsing comic magazines. But that evening, behind the information desk in front of the poetry section, stood Carrie, a swan with green eyes, in a dress and lipstick. "Blimey . . . welcome the roadworks," thought Aidan. "I'm giving in to books!" And he did. That route became the most awaited part of the day. Every weekday between five and seven, depending on when he left the office, he would return in the hope of finding Carrie. Uninhibited, on his second visit he asked her name. He started to pretend he was interested in poetry by asking random questions about writers and their lives, and Carrie pretended she hadn't realized this was a front.

If Carrie happened not to be around when Aidan was in transit, he would wait, browsing a volume or two. Sometimes Carrie hid behind shelves to test how long he would wait.

"All right, I have a proposal," said Aidan, after almost a month of flirting.

"What is it?"

"If tomorrow we meet each other here again, we'll go for a drink together somewhere."

"Deal!"

And the day after, Carrie didn't leave the desk for a moment, not even to go to the toilet. It was summer, and Aidan showed up holding two ice-cream cones, one in each hand, leaving a sticky trail behind him.

Time, what a funny detail, Aidan thinks to himself today. One child, two mortgages later, illness, their own book business, a few big fights, but above all a lot of laughter, the two are close to retirement and still counting the blessings of their undying union. When asked, they can't tell what has kept them together all this time.

Along the same streets, there are roadworks again, only bigger, and Aidan wishes he could be more forgiving of them. Entire strips of ground are fractured, tossed and strewn about. Week after week, layer after layer, corner after corner, scrapers and hammers skin the crust of Soho, rummage its guts, and expose its veins, the old water pipes that keep bursting and need to be replaced. Their twenty-first: a new leap second will come again between them at the stroke of midnight. A souvenir of the past or a robbery of the future. Aidan remembers how experts' opinions and explanations had filled the news, and that he and Carrie had talked about the leap second with wonder. Adding a second was all a matter of matching clocks to planets, as it were, to smooth a crease between man-made time and the larger astronomical order. The time we see displayed on computers ticks exactly to make the sum of day and night last for twenty-four hours made of a total of 86,400 identical seconds. But even if we don't perceive it, Earth occasionally takes longer than that to spin around itself. Its internal wiggles and diggings, because of earthquakes or other geological melts, cause its rotation to slow down. The ocean tides, pulled by the gravity of the moon, impede it too. When in need to make up for these unpredictable delays, scientists agree to lengthen the year by a second.

On their first leap second, in 1973, at the end of a long day at work, Aidan hoped to meet Carrie at the bookstore. Today she is baking their celebratory cake in their home.

Now in front of the Goodge Street Tube station, Aidan smiles sweetly because sweet is the memory of what happened underground twenty-one leap seconds ago.

*   *   *

In a heartfelt letter to his friend Friedrich, the German poet Rainer Maria Rilke shares his latest conclusion on the operations of love: "I learned over and over that there is scarcely anything more difficult than to love one another. That it is work, day labor, Friedrich, day labor; God knows there is no other word for it."

Intimacy is not an immutable talent. Rather it is a journey. Like other dexterities, it hones itself by trial and error. Intimacy means to enact, rehearse, and polish modes of connection. Each time we begin to establish a relationship, we have a chance to *learn* how to be intimate, in both the short and the long term. Whether the union lasts for a few months, decades, or a lifetime, the refinement requires time.

Time is a dimension intrinsic to the exploration of the mind, but how the mind, or the brain, keeps track of time, and how we subjectively perceive it in a broad range of events and situations is still an open question. Since the 1800s, studies of mental chronometry have assessed the time course of mental events. These studies have mostly concentrated on measuring variation in people's reaction times in the context of cognitive tasks of various degrees of difficulty. The consideration of a temporal dimension in the unwinding of interpersonal and emotional routines is equally crucial but is a relatively recent addition to laboratory investigation. Techniques such as functional magnetic resonance imaging (fMRI for short) have extensively been employed to map emotions on the geography of the brain. However, there is not a one-to-one correspondence between a given emotion and one area in the brain. The overall activity connected to each emotion is shared across networks of regions that work in parallel. One region may be more prominently engaged in a particular emotion, but at the same time will assist and underlie the processing of others. In this mesh of networks, coordinated time dynamics is paramount. What counts is not only the magnitude of an emotion but also its temporal width, or how long it takes to reach or lose a certain degree of its intensity.

Our emotional life and our social interactions unfold across several timescales. When interacting with others, we observe, we perceive, we act, we remember, we imitate, we share, and we forget. Possibly, we adopt, change, or abolish habits. Things happen in the order of submilliseconds, minutes, hours, days, and weeks. Even months and years. From the oscillation of neural waves to a rush of electricity across nerves, from the expulsion of neurochemicals to the acceleration of breathing, and from the proliferation of neurons to epigenetic tagging on DNA, the physiology of togetherness plays out at different speeds. It takes less than an instant to scan or mimic a facial expression. Angry or joyous reactions can peak in a matter of seconds. But for our mood to mount or sink, it may take a few minutes. The abolition of an undesired automated behavior may take years. We could ask: How long does it take to really get to know someone?

Every individual has their own timescale, as part of his or her emotional style, which is in turn an outcome of genetic composition and biographical experience. In a pair, these two scales will meet, compare, and adjust. This involves comprehension and acts of will. Aidan is faster than Carrie at initiating or recovering from an argument. While he forgets easily, Carrie mulls over details for days. On the other hand, Carrie is fast at picking up his mood, while Aidan occasionally needs explanations for her words or actions. In an effort to uncover individual differences in emotional style and reactivity, research has begun to look in detail at temporal dynamics, also in connection to distinct personality traits or conditions. For instance, neurotics—who have a tendency toward negative emotionality and to persevere in it—take longer to recover from the spasm of an upsetting event. The part of the brain that engages in the action of coping with that event—the amygdala, to be precise—shuts down more slowly. Similarly, when we are depressed our amygdala may take ten or fifteen extra seconds to recover from the effect of being exposed to words related to a negative mood.

We may wonder how long it takes, then, to build intimacy. The answer to this question will greatly depend both on the degree of closeness and the individuals involved. A few studies have attempted to artificially instigate intimacy. From the swapping of secrets and eye gazing to self-disclosure and expression of defenselessness, techniques have emerged to help spark as well as sustain closeness over time. For instance, a simple physical exercise that involves letting oneself fall backward into the grip of a partner's arms helps nurture feelings of vulnerability and trust, both essential ingredients in the creation of intimacy.

One of the first studies testing whether closeness can actually be generated among strangers dates back to 1997. Two groups of participants (in this case all heterosexual) were divided into pairs and asked to entertain each other for about forty-five minutes. One group of pairs engaged in small talk. The pairs in the other group asked each other questions that increasingly exposed personal details about themselves and were supposed to facilitate intimacy and mutual knowledge. The personal questions regarded a person's qualities, wishes, aspirations, values or secrets, and embarrassing facts about themselves, as well as information about their relationships with their families. For instance: "For what in your life do you feel most grateful?" or "How close and warm is your family?" and "When did you last cry in front of another person?" At the end of the procedure, when the researchers measured levels of closeness between participants in both groups, the pairs who traded the more personal details declared higher levels of closeness. In a similar, more recent experiment randomly matched speed daters were asked to fake that they were already in love with each other. The general idea behind the role-play was that simulating an emotion has the power to make it stick. Like in the previous experiment, the participants shared personal details and secrets. With the pretext of practicing palm reading, the fake couples were also encouraged to touch each other's hands. In addition, they stared attentively into each other's eyes, an action that is known to be more powerful at establishing a connection than just looking more broadly at a face. At the end, 45 percent of those who participated expressed the desire to meet again.

Another experiment looked at the neural substrates that underlined first-glance judgments of potential partners who met at a speed-dating event. The outcome of the study revealed that keenness to pursue someone could be mapped to brain activity into two distinct portions of a region of the brain called the dorsomedial prefrontal cortex, one assessing physical attractiveness, and the other psychological compatibility.

Although a five-minute or hour-long interaction in an experimental setting may induce a certain level of closeness, that intimacy is not comparable to the loyalty and dependence that arise from an extended shared history of behavioral and emotional familiarity. Some of the techniques used to generate intimacy have been tested within relationships to improve them, confirming the possibility, as it happens for instance in arranged marriages, that commitment, understanding, and accommodation can grow between people who do not begin with a passionate foundation. As Rilke would say, the whole endeavor takes effort.

* * *

A bouquet of violets in hand, Aidan turns right to walk to the hospital on the corner, where his son Anton works. Anton decided to become a children's doctor after having survived leukemia. He had just been born when Aidan and Carrie fetched him from an orphanage in Kiev. During the time of Anton's illness, Aidan and Carrie faced the most important decisions of their lives. They were little more than a boy and a girl when they heard the diagnosis, but they had to grow up fast to gather all the wisdom they needed. Defenseless but all the same defiant. The tears welled when the doctors disclosed the positive prognosis and their future looked bright again. After that trial, choice after choice, there was no challenge they could not endure. Danger has the power to bond. So does sacrifice.

Relationships mature. Just like other aspects of our personality, attachment shows a temporal development that is mirrored by behavioral changes and underlined by physiological alterations. The euphoric sickness of a love-struck couple experiencing butterflies in the stomach differs from the contentment and trust enjoyed by partners in a long-established domestic nest. Decades of research have identified a few distinct molecules as the drivers behind specific phases of love. Things are definitely not black and white, but it's fair to say that the neurotransmitter dopamine governs the early blind-and-obsessive passion in romantic love, while oxytocin and vasopressin favor long-term commitment. A few important words of clarification: No single molecule equates to, or is responsible for, a particular emotion, mood, or inclination, let alone a loving style. The manifestations of our feelings and our dispositions to intimacy result from the ongoing concerted action of countless parallel events in our bodies and of elaborate encounters and fluctuations of neurochemicals in the brain that rise and drop, disperse and concentrate, as we dart from one state of mind to another.

Dopamine is known to awaken desire. It brings excitement, motivation. It is released in the reward center of the brain, an ancient and essential device that we share with other living creatures, from bees to elephants, and that makes us experience and remember the pleasure of anticipating enjoyable rewards, including food and sex. Tiny in size, oxytocin and vasopressin are instrumental for several aspects of social behavior. They circulate in the brain, and via the bloodstream they also reach organs such as the heart and intestines, as well as the uterus in women and testes in men. The name oxytocin is derived from a Greek word that literally means "sudden delivery." When giving birth and during breastfeeding, mothers produce large amounts of oxytocin, which also works to cement their bonding to the newborn. Oxytocin is also released during female orgasm. Vasopressin, which is derived from the Latin words for "vessel" and "pressure," peaks in men during ejaculation. Oxytocin and vasopressin can help reduce stress and anxiety, and promote trust and social bonding.

A lot of what we know about their role in pair mating has emerged from observing the behavior of two particular kinds of field mice: mountain and prairie voles. The former are promiscuous, the latter monogamous. What contributes to making the prairie voles faithful is a higher number of oxytocin and vasopressin receptors in their brains. More oxytocin makes females attach to their partners. In males, vasopressin inspires possessiveness of their partners as well as territoriality and aggressiveness in defense of offspring.

Now, we can't say that dopamine or oxytocin and vasopressin are exclusively present or play undivided roles in passionate lust or committed attachment. Not confined to the initial stages of romantic desire, dopamine is produced each time we encounter novelty or anticipate something—say we are waiting for the arrival of a loved one at the airport—even after thirty-five years of marriage. Oxytocin and vasopressin also belong to early romance. One study has shown that just a few months after falling in love, couples have a lot of oxytocin in their blood, in levels that are even higher than those found in new parents. Six months later, those levels were found not to have dropped. Instead, they stabilized and the higher the original amount of oxytocin, the more pronounced was the degree of reciprocity in the couples. Evidence suggests that oxytocin and vasopressin facilitate qualities, and affiliated behavioral traits, such as trust, empathy, generosity, and dialogue, which become especially handy later in a relationship, in the mature stages of commitment. For instance, heterosexual couples who took oxytocin before discussing thorny issues such as finances or free time, spent more time looking into each other's eyes, validating and opening up to one another rather than showing defensiveness, criticism, or scorn during their argument, at the end of which even their levels of cortisol, an indicator of stress, sank.

The apparent neurochemical fluctuations that distinguish the evolvement from an early young to a more mature phase of attachment becomes extremely interesting with regard to the functioning of our internal clock and our subjective experience of time.

Built on experience and memory, our internal clock helps us predict and estimate the duration of time. As it happens with many of our faculties, with age the precision of our internal clock declines. In particular, what becomes more pronounced as we get older is the compression of time, or the perception that the duration of a given period of time is shorter than it actually is. This link between the advancement of age and the loss of precision in our internal clock has been proven by a simple and elegant experiment. Participants of different age groups, from twenty to around sixty years old, were asked to wait for three minutes and then say when, for them, the time was up. The youngsters were pretty accurate in their guesses, which were on average off the mark by three seconds at most. The estimates of the older group were more consistently wrong. Those in their sixties announced the three minutes were up after about three minutes and forty seconds, indicating that their subjective clock ticks slowly and that time seems to run faster for them.

There is now an increasing amount of evidence that this change is linked to a decline in dopaminergic function. The metabolism of dopamine plays a role in the ticking of our internal clock. If you inhibit the dopamine system, estimates of time will go awry. When we are young budding lovers, it's easy to say, "I want to be with you forever." As we age, time speeds up and we become aware that that forever is actually nearer. In old age, a life spent together feels like an instant. Instead of forever, we are more likely to say, "It seems like yesterday that I first met you." In Aidan and Carrie's case: only twenty-one leap seconds.

*   *   *

You can tell what somebody is like, the proverb goes, by the company they keep.

Being continuously exposed to another person, sharing a past and witnessing life with them, makes us adopt their ways and possibly their ideas, opinions, and worldviews, sometimes with generosity and compromise. The unavoidable reciprocal entrainment, both physical and mental, accumulates on a moment-to-moment basis, forming a sort of sediment over time. Hearts beat in unison, respiration patterns keep up with one another. Facial expressions are mirrored. Posture is mimicked. Gaits align. On the whole, behavior is replicated. And brains synchronize.

Aidan's occasional restlessness has been mitigated by Carrie's effortless composure. Over the years, Aidan was taught the pleasure of choosing and sitting down to read a book. By reflection, Carrie's fuzzy sense of commerce improved by watching Aidan balance the business's orders and bills. Together, as they raised Anton, they learned how to unify their parenting views and become a single harmonious voice for their child, both when they needed to encourage his conduct and when they needed to scold it.

By and large, synchrony catalyzes bonding. In one study, pairs of individuals felt mutual connection when their response times in an interaction game harmonized. All they had to do was tap on a keyboard at a given regular frequency. In turn, their partners were instructed to respond with a tap before the next tap could be received from their coplayer. Beat after beat, and by trial and error, the players entered a harmonic rhythm of communication. Each member of the pair then rated things like how much they liked or felt close to their partner, how much they trusted them, or if they would like to work with them. A stronger experience of synchrony in the game corresponded to more pronounced affiliation.

Sharing emotions helps us converge with others and marks how we experience the world. One set of studies has explored how we react similarly to emotions that unfold before our eyes in film. Aidan and Carrie loved going to the movies together, and they enjoyed sharing moments of exaltation, terror, and tenderness. They made a point to go every fortnight, if they could, and never miss a new release. Traditionally, the first day of the year was cinema day. One study has shown that people who watched a thirty-minute clip from Sergio Leone's memorable western *The Good, the Bad and the Ugly* displayed synchronized brain activity, not only in areas governing audio-visual sensation, which one would expect to be at work by default in any shared viewing, but also in areas that process emotions, as well as in areas responsible for mental simulation. Peaks of global activity across the entire brain cortex matched moments of heightened emotional intensity in scenes with unexpected plot turns or gunshots and explosions. Strikingly, a closer inspection that dissected the clip into discrete segments with a predominance of specific elements, objects, or actions detected more selective synchronous activity in areas that are specifically at work when we, for instance, concentrate on faces, buildings, or hand-related movements. In a similar study, participants were twice exposed to a clip from the television series *Desperate Housewives* and asked to watch it one time from the perspective of a detective and another time from the perspective of an interior decorator. Measurements of brain activity across viewers revealed synchronization in response to the episode, and this was stronger when the viewers adopted the same perspective, suggesting that the alignment of brain activity may contribute to the adoption of a similar worldview.

* * *

Language is a synchronizer. Spoken emotional words, for instance as narrated through stories, are particularly powerful at inducing attunement. One study showed that narratives told to a group of people elicit synchronous activity in auditory, linguistic, and emotion-related brain circuits, the latter more active if the narration included more negative emotions, such as fear or anger. In addition, the synchronicity corresponded to enhanced connectivity between those areas.

Aidan is a talker and knows how to tell a good story. He likes to make things up and to leave Carrie guessing if what he said is true. If Carrie happens to enjoy something he says, he is in heaven. However, for a while at the beginning of their relationship, they experienced a few bumps in their communication when trying to reach a balanced level of expansiveness. Often Aidan overdid it with his declaration of affection. He left little cards for her in which he openly said romantic things like "I love you" or "I can't tell you how much I care for you," "I miss you," etc. He also spoke those sentences on a regular basis. Carrie knew he loved her and felt there was no need to put things in black and white. Aidan needed to express his feelings. She was much more at ease when she and Aidan simply shared and talked about the most mundane things. There is a thin line between what lovers wish or expect to hear from each other, between the tacit and the outspoken, the essential and the redundant. Too much kindness can be mistaken for unnecessary formality. Laconic speech can be mistaken for lack of warmth. Over time, they reached a compromise. Aidan reserves his wordy outpouring of devotion for rare occasions, which together they call the "shallow moments," and he introduces it with a bow.

Relying on habit and exercise, our brains are in general good at making predictions. This holds true also during a conversation. If we are well acquainted with someone's speech, we will easily predict the words they'll use to express themselves and how they will articulate their thoughts. That's how we learn to finish other people's sentences.

Two people who communicate in the same language and talk to each other on a regular basis will share vocabulary and will accumulate knowledge of each other's speaking habits, as well as word choice and syntax preferences. This knowledge shapes the temporal profiles of their brain responses to each other's speech and helps their brains lock. When we listen to someone else speak, we absorb what they say by a two-stage process. One is anticipation; the other is perception. The former influences the latter. Even before someone speaks, we have a hunch of what they might say. This prediction phase influences how we perceive what is actually said and the degree of predictability of their speech enhances the overall neural synchronization. Imagine, for instance, that someone is about to describe a simple action to you. If that action is explainable in highly predictable words, say, "a man is fishing on a boat," attention among listeners will be high before the sentence is uttered, as well as during the perception phase. During both fragments of communication, the temporal profiles of brain activity in the speaker and the listener will align, especially in areas that are involved in lexical-semantic processing as well as prediction. So if we have knowledge of the words they would be using in a given situation, our brain would synchronize strongly.

Conversation is not made of words alone. Silence counts as much as talk. Nested in the rhythm of a conversation are gaps that sustain its structure as much as the fullness of its spoken components. Basically, when we talk we take turns, and we do this without realizing it. This habit is universal across cultures and older than language itself. Just think of the turn taking in vocalizations that takes place between infants and caregivers, and there is evidence of it taking place among primates too. Turn taking is a skill we have honed since childhood. It's in place so we can give others a chance to speak, so overlap between sentences is reduced, and so people don't talk over each other. Overall, the art of taking turns involves work. Every turn we take lasts on average two seconds. A typical gap goes for about two hundred milliseconds, the time it takes to pronounce a single syllable. For the turn switching to be efficient, we, in incredibly short intervals of time, predict, encode, and comprehend what we hear and get ready to produce our own sentences. It takes between six hundred and fifteen hundred milliseconds to prepare for language production, depending on whether we are to emit a word or a short sentence. We already plan our next turn while the interlocutor's is still ongoing. Knowing how keen Aidan is on talking, to educate him to be a better partner in conversation, especially with strangers, together they came up with a trick. When he didn't realize he was talking over people, Carrie would say: "Oh, did you all hear that sound?"

* * *

Timing is everything, it is said. It was quarter to midnight on December 31, 1973. London was united in anticipation. How odd that at the end of the year we are obsessed with marking borders—of life, of change—that on any other day most probably would go unnoticed. During those final hours, minutes, and seconds of a year, the course of time becomes especially visible.

In his own drift toward the future, Aidan was ahead of himself but impeded by a knot he was impatient to untie before everything dissolved into the big celebrations. He was full of hope but felt the itch of something left undone.

After work, he had stopped at the bookstore to meet Carrie just before closing time and wish her a happy New Year. During his entire Christmas holiday, spent at his parents' place, he only thought of Carrie. He remembered she said she would work the evening shift on New Year's Eve, and they agreed he would drop by to wish her a happy New Year. Time was ripe, he had decided, to let her understand one way or another that he really liked her, that what he felt was serious. He had no idea how he was going to do it, what he was going to tell her. At least he would suggest, if she had no other plans, she join him to watch the fireworks from Primrose Hill and perhaps meet the next day to see a movie together.

It was time for them to take a leap.

But, strangely, she wasn't behind her desk at the bookshop, and her colleagues didn't know where to find her. Aidan waited outside for a short while, peeked in again, then went back outside, looking out for her down the street. Fifteen, twenty, thirty minutes went by. And then the shop closed.

He should have invited her before. Disappointed, Aidan strolled along the Thames for a couple of hours, had fish and chips for dinner, and then decided to walk back home. On the streets, he overheard a few people talking about the second that was to be added to time. Everything was going to be in better agreement with the motion of the earth.

When he was a child, he was taught how to time the end of the day. As mirrored on his much-loved first wristwatch—which had belonged to his granddad and figured the phases of the moon—he knew it was time for him to go to bed strictly before the short dial aligned with his forearm. In winter, that came after soup and four pages of *The Wonderful Wizard of Oz*; in summer, when all the four o'clock flowers in the garden had opened and after he had collected in a small blue velvet case at least a dozen of their large, wrinkled black seeds.

What would a second matter to Aidan's expectations for the day, let alone for the year to come? What could it bring?

Well, he only desired one thing. Aidan slowed down his pace. He was captured by the day's tide, slowing to catch up with the planet. When he reached Centre Point, tired of being on his own, he changed his mind and decided to hop on the Tube and go see the fireworks. The train was packed. Everyone except him belonged to a group. A girl in sandals made space for him next to her.

"You can't be on your own on New Year's," she said.

"Stay with us," cheered the tallest guy in the group, visibly drunk.

They all seemed comfortably in tune with the rhythm of the night, and Aidan was kind of reassured that they knew better than he did how he was supposed to spend the end of the year.

Ten.

Nine.

Eight.

Aidan didn't count along. The train stopped at Goodge Street station and the doors opened.

Seven.

Six.

In a bright red coat, a couple of books under her one arm, wearing makeup, and with her hair in a bundle, Carrie emerged from the platform, the doors closing behind her.

A bottle of champagne popped too early.

Five.

Four.

"I came to the shop, but I couldn't find you."

"Doesn't matter, Aidan. We are both here now."

Another bottle popped open.

Three.

Two.

One.

"One again!" someone yelled.

Aidan shivered. That was a moment to condense in one impulse an intention he had been barely able to contain. They both leaned and tilted.

A kiss is not a trivial enterprise. Nerve endings on the face brace. Muscles around the mouth pull the lips on their corners and make them wrinkle. Pupils dilate, eyes close, breathing deepens. A kiss is a dot that starts a line. On average, we tend to remember the details of a first kiss better than we remember our first sexual experience.

Through the cheers, shouts, and whistles of the passengers, New Year's wishes from the driver permeated the carriage through the speakers.

Life stuttered at the turn of the year, the leap second lodged somewhere between old and new, between now and forever. Between a life of their own and a life shared.

Transit of Venus

The first scene is over. Next comes four minutes in the dark. In the wings, Ryan's heartbeat is irregular. A soft vibration startles him. He forgot to leave his mobile phone in the changing room, and it's now pressing on his left thigh.

In the audience. Meet me at the back after the show. V

Ryan expected Vanessa, but not until the next day. His reaction is inscrutable, even to himself. Well, no need to find a fresh excuse for Laura, his wife. After the show he's supposed to attend the usual party for the premiere, the kind of event she stopped escorting him to a long time ago. He'll be glad to disappear from it early.

Vanessa first came between them eight years ago, on the other side of the world, when Ryan and Laura were not yet married but getting along well. Vanessa had shown up to a dress rehearsal. She was a friend of the director, and she was interested in studying actors on-stage for her doctoral dissertation on authenticity in performance.

"Could I interview you?" she asked. "In the name of knowledge," she hastened to add, the remark a blunt attempt to hide her immediate attraction.

Her auburn hair and her scent of clean peaches and honey, mixed with an ability to talk about abstract stuff without sounding sophisticated, had made an impression on him. Then, she was single. Ryan and Laura had already been together for two years, a period that straddled their final stint in the academy and their early attempts to launch themselves into the acting world. They had no recipe, but they both claimed a desire to be famous. Laura was in awe of Ryan's talent, and because of it, she accommodated his neurotic reactions to failure. Secretly she wished him a future as a film star, while she would have been content with a small job in a local theater.

The big production traveled, and for two months Vanessa followed the company wherever it went, on a tour across the world, to gather data. Fearless, she stared, she asked questions, she took notes, she wrote, she seduced, and she imagined. She dared. She became inevitable. They fucked on the last stop of the tour, at night, between rows at the back of the orchestra, the theater shut and dark, just days after Ryan had asked Laura over the phone if she would marry him. Both Ryan and Vanessa registered a seismic shift in their overall sexual tenor, in latitude and depth, temperature and flow, like they had migrated over an entire continent on a short flight and back. Until then, Laura had been a comfort. Vanessa stirred that. For Vanessa, Ryan was an amplification of her pleasure allowance, even his shoulders wide enough that she disappeared under him, and without feeling it was wrong.

Neither of them dared ask or explain why it happened. Even less were they interested in the "how long?" Good-bye. Good-bye.

"I'll send you the transcripts of the interviews. Please tell me if there is something you'd rather I omit."

"Will do."

Only later, after the tour was over and the two found themselves geographically apart, did the deed itself resonate and raise doubts. There were questions of metrics and scale. The minuscule confounded with the gigantic, imprecise prospects rushed into view.

Ryan initiated a dense correspondence that held them together for a whole summer, at the end of which they saw each other again, thanks to one of Vanessa's research trips. Then came Ryan's marriage. That didn't stop it. While she finished her PhD, Vanessa found a boyfriend, Kevin, or Lord Kevin, as everyone called him because of his outmoded manners. They met again, this time when Ryan performed in her city. Laura became pregnant with his first son. That didn't stop it. Vanessa had a daughter a year later—from Kevin, she thinks. That didn't stop it either. Vanessa published her research, but her fascination for the boards didn't fade.

Their carnal and epistolary exchanges grew to be a mutual guide, a ramp for flights of evasion, as well as a landing site for emergencies.

Okay. Wait for me outside. R

His second entrance is lighter. What Ryan has always loved about being an actor is the privilege of deception. He relishes the chance of starting all over every time he goes on stage. He can divest himself of everything he wishes to ignore. Sometimes it was the prospect of aging, his frail parents, a place in the local kindergarten for his son, the lack or abundance of fame. Sometimes it was Laura or Vanessa or himself. Onstage he always soared above the labyrinths he had weaved for himself. Vanessa is attracted to this and envies it. Standing outside now, with her back to the stage door, she wants to be surprised or to hear him call her name. It'll take him a good twenty minutes to come out, she reckons. They are both accustomed to the trepidation. In a strange act of penance, Ryan makes her wait by letting everyone else in the cast use the shower first.

* * *

Natural or cultural. Unavoidable or deliberate. Innocent or sinful. Legitimate or unacceptable. Infidelity, and whether we are wired for monogamy or polygamy, is a huge question, an old crease between the laws of human nature and norms of social order that is beyond the scope of this chapter. Whatever side we wish to take on the issue, one thing is certain: infidelity happens, and extensively. Another fact is that only a small fraction of animals are exclusively monogamous. Eagles, for instance, and albatrosses. Swans, for whom the job of migration, nest building, incubation, and rearing is so time-consuming that straying would considerably affect their resources, are monogamous, though researchers can't agree on them either. Around less than 5 percent of all mammals do pair bond. The prairie voles mentioned in the previous chapter are among the members of this small elite. So are wolves. The rest appear to be largely promiscuous. Bonobos, for instance, are renowned for their unbridled and jealousy-free sexual habits. We humans are definitely not exempt from acts of promiscuity. In one of the two Kinsey reports, comprehensive studies of sexual behavior conducted in the 1950s, 36 percent of husbands and 25 percent of wives reported unfaithful behavior. More recent studies confirm and exceed these figures. Infidelity is also a major cause of divorce. Men and women have committed homicides in retaliation for infidelity or to eliminate the spouse who opposed it.

The infidelity dispute raises long-debated evolutionary distinctions between males and females, with frequent invitations to draw parallels between us and our prehistoric predecessors. According to basic Darwinian evolutionary theory, a species' and an individual's traits and features, both physical and behavioral, are advantageous if they also provide reproductive success; that is, if they don't interfere with the generation of offspring that are also capable of surviving and reproducing. In his short essay "On the Metaphysics of Love," an interesting read for boys and girls in the dating jungle, the philosopher Arthur Schopenhauer says it in no uncertain terms: "The state of being in love, though it may pose as ethereal, is rooted in the sexual impulse alone; indeed, it is only a more closely determined, specialized, and (in the strictest sense) individualized sexual impulse." Heterosexual individuals are unconsciously under considerable pressure to engage in kinds of behavior that encourage their reproductive success. In Schopenhauer's words, that sexual impulse is "the strongest and most active of all motives, and continuously claims half the energies and thoughts of the younger part of mankind."

In following this impulse, the two genders present differences in their approach to infidelity. For a man, sleeping around guarantees reproductive success because by making several women pregnant he will have the opportunity to expand the spread of his genetic material. Depending on circumstances, men may also have more time than women to find sexual partners because their parental investment may be less time-consuming. For a woman, on the other hand, straying is less advantageous because once she has found a partner who can make her pregnant, there is no pressure to find another one, unless there are prospects of finding a considerably fitter and more beneficial partner in the long term. Along these lines, men and women would also display different kinds of jealousy. In a famous study, participants, mostly college students, were asked to imagine their partners with someone else and to ponder whether they would be more upset at the idea of their partner involved in a liaison that just involved sex or one in which their partner had formed a relationship that was deeply emotional but involved no sex. The results were revealing: 60 percent of men but less than 20 percent of women were more distressed by sexual infidelity. Evolutionary psychologists find the motive for these results in the concept of parental certainty. Unless the resemblance is undeniable or the fact is made certain with a paternity test—and those were certainly not available in prehistoric times—men can never be certain of a child's biological paternity. This would make them in general more jealous of a woman's sexual infidelity. By contrast, women would be forgiving of a sexual fling but more fearful about an emotional involvement because the latter might motivate their man to abandon them and their shared offspring for a new lady. The distinction of sexual jealousy for men and emotional jealousy for women has become a stereotype that contributes to the propagation of contradictory standards in attitudes and behavior in many cultures. The assumption being that men are more sexually irrepressible than women. In light of their evolutionary drives and the moral embargoes on adultery, both men and women would seem to be caught in a trap. A man would make sure to have children with his wife, but then go on to sleep with, and possibly make pregnant, as many women as possible, to spread his genes. In turn, a woman would first settle with a rich provider of resources for her offspring and then, in search of a genetically fitter companion, stray, for instance, with her tennis instructor.

As a species, we definitely have the capacity to nurture feelings of exclusive attachment. Monogamy has evolved to facilitate the rearing of young in communities. Some people are unfaithful while cherishing and maintaining a monogamous bond.

In their book *Sex at Dawn*, authors Christopher Ryan and Cacilda Jethá have provided challenging evidence against monogamy. They argue that valid evolutionary explanations favor polygamy for a successful social organization. Ryan and Jethá say that a regimen of overlapping sexual relationships, or frequent SEEx, "socio-erotic exchanges," as they name them, must have strengthened the networks of even the most ancient foraging societies by providing fruitful cooperation, affiliation, and reciprocal gratitude. Sex and love can often be mistaken for each other. But just as they can coincide, they can also be distinct, and sex may have functions beyond the achievement of pleasure, and work as a lubricant for social cohesion.

Ryan and Jethá claim that fluidity in sexual relationships must have ensured wide fertility and social stability for generations precisely by forming groups with children of dubious and collective paternity. Monogamy and the institutions, such as marriage, that endorse it are regarded as a structural thread for the social fabric. Marriage, however, is a social construct, and not a reflection of human nature, or at least not its exclusive fate. While the sanctioning of monogamy has been historically robust, there are still viable communities for which polygamy has been and continues to be the norm. In villages in rural Mozambique, a total count of one hundred and forty men entertained relationships with hundreds of women at the same time. Up to one hundred were formal wives, more than two hundred were long-term sexual partners, and another couple of hundred were occasional partners, with each man being involved in around four sexual liaisons at a time. In the Trobriand Islands there is a tradition for young women to incite men to have sex and to punish them if they don't oblige— suggesting that women's presumed reticence about sex is culturally imposed. Eighteenth-century travel reports from Tahiti confirm various forms of licentious behavior among local women. There is historical evidence that in Iron Age Britain men, especially brothers, shared wives, and, of course, there are endless accounts of sexual promiscuity from the Dionysian festivals in ancient Greece.

In other words, while monogamy and marriage are one way of experiencing sexuality, the urge to sexual promiscuity seems impossible to sedate. If it wasn't for man-made opposing injunctions, a looser sexual liberty is something people might embrace more openly, in tune with our close cousins, the bonobos. Though, the sense of the forbidden is perhaps an equally strong catalyzer for it.

*   *   *

"Do you believe in accidents?" Vanessa asked Ryan in one of her letters, which she always wrote on rice paper. "I don't," she hastened to add for herself. "How could we have not come into each other's lives?" A tension between unruliness and order, and between confusion and focus, obsessed her. She sent letters to him at the theater address, each disguised as something else—a bill, an official invitation, or some promotional junk. He sent his to the university department where she worked. They enjoyed the old-fashioned way of communicating, in addition to the alacrity of e-mail and texts. Slow, yes, but also a lot safer. Vanessa needed to be particularly careful because she was clumsy with technology. Her messages might have ended with "forever yours." She always wrote to Ryan on impulse, uncontrolled, now and then behind the shield of superior forces—"Today even the air suggests I get in touch"—although she regularly asked him if her immoderate confessions of love were sometimes too much to take, making it absolutely clear that he, on the other hand, was never intruding. "I'm an actor. I can deal with drama. Remember?" Especially if it came from her. Vanessa would describe to him the overwhelming number of projects she worked on and couldn't let go of, how she sometimes struggled to make them intersect. Her letters to him pursued an acute desire to coincide, not just in physical attraction, as it all started, but in broader intellectual affinity, to the point that she would sometimes belittle her scholarly identity and aspire to be an artist herself: "You know," she wrote, "I've come to realize that, after all, what you and I do is strikingly similar. Research is like theater. You perform and I declaim. We both search and find. Only, I wish I knew better how to lie in order to tell the truth." And with a similar longing, Ryan responded accordingly: "I read somewhere that the words 'theory' and 'theater' might have the same etymological root. Did you know that?" "You see, you are beginning to do my job!" Vanessa said.

Occasionally, Vanessa would feel confined and her gasps for fresh air would become louder. "He's being particularly difficult at the moment," she might say, speaking of Kevin, who according to her was prematurely painting a too bleak view on life: his new high-responsibility post, the future of their child, their receding savings, his rampant baldness, the unforgiving professional competition. Dispirited, she would nevertheless always be encouraging: "All this belongs to life, Kevin. Get over it." But in reality Kevin would do absolutely nothing to get out of his rut, and she detested his resignation. In contrast, Vanessa loved the way Ryan always grasped her need for idealism and impracticality, the way he gave her space to believe there was always a chance to be young and room for change. "Freshly yours" or "erratically yours" was how he signed his messages. With Ryan, her hunger for the uncharted was quenched. For each other, they reserved the better versions of themselves. Laura did settle for a low-key job in a local theater. Ryan's success was superior but never enough for him, and he vented his frustration on his wife. With Vanessa, on the other hand, he wasn't as self-obsessed. He was kinder. He didn't feel the urge to impress.

Over time, Ryan and Vanessa found themselves trading recipes for companionship and parenthood. "My son is sleeping, and I wish he was still up because he saves me from thinking too much. Does this happen to you?" asked Ryan. "Never miss an opportunity to be gentle to your wife," advised Vanessa. "Teach her how to be just as brave as you are," said Ryan, referring to Vanessa's daughter. The mature choice of mixing family affairs with their own entanglement may have felt weird to a lot of people, but they couldn't help it, and they were good at it.

For the first couple of years, their union went entirely unchallenged, by them and by those around them. The director who introduced them was the only person Ryan had spoken to about his affair. Vanessa had only told her closest friend and her mother, neither of whom would ever challenge the arrangement. As far as Ryan could tell, Laura didn't suspect anything. Or at least she never voiced any jealousy. One Christmas, three years into the affair, Kevin confronted Vanessa with his suspicion when, the house full of friends and family, she was completely absent, forgot to buy him a present, and burned the turkey in the oven. He asked her why, since her research was finished, she had to continue to see Ryan, and where were all the other actors and actresses she might interview?

"Come on, this is a lifetime project. It's thanks to my work with him that I got my tenured position. Plus, you know, he's really better than the others."

Vanessa and Ryan didn't actually need the inquisitiveness of their respective partners to occasionally reconsider their actions. They regularly generated doubts for themselves. That was when the two would withdraw, in turns. Chased by guilt and insecurities, they imposed on themselves a regimen of resistance and abstinence. They could go for a couple of months without seeing or hearing from each other.

"We will find a way to direct this force in the best direction," she said optimistically. "We need to," he said, "for everyone's sake." They would have to endure in scarcity, in separation, in occasional exaltation. Once Ryan signaled that he wanted to end it: "We have exchanged wisdom and space, my dear. We were acrobats, always soaring and landing with skill," he wrote. "My life compass has become sharper thanks to you, but now is time for us to part." "No, why? We can bear this," she said. "Your wings are strong," he said, "and the winds are in your favor. We'll be fine. For aspiring souls, the wind of dreams is always the most prevailing."

* * *

When Vanessa was a teenager, her parents divorced after her mother discovered her husband having sex in the garage on the Fourth of July. After that episode, Vanessa promised herself never to marry. She also thought she would never betray her own partner. Ryan's parents were still together, although his father's occasional straying episodes were a secret neither to him nor to his mother and his brothers. Everyone simply turned a blind eye. At a big family dinner, drunk and in the presence of a bunch of relatives, Ryan's father once said there must have been a protein, which he called "cuckoldin," or better an enzyme, "What do they call those things?" that made men irrepressibly prone to sleep with other women. In search of justifications for their conduct, those who betray are prepared to go to great lengths. As we have seen, biology is an attractive and legitimate authority dispensing clemency to those who feel the responsibility of their actions. And so the pleas pour in: My brain made me do it. I am wired to cheat. It's the hormones. I've got the infidelity gene. All thick shields against the threat of moral accusations.

Levels of the hormone testosterone are connected to sex drive, and men with higher levels are in general more likely to cheat. Testosterone levels typically decline with age and in men who are in committed relationships but remain high in single men. Interestingly, testosterone returns to high levels in men who are in relationships when they cheat or if they show interest in cheating, suggesting that the hormone helps men respond to mating opportunities.

The occurrence of the vast majority of complex traits and behavioral outcomes—with the exception of proven monogenetic conditions such as Huntington's disease—is mired in a complex mesh of intertwined genetic and environmental contributions. Infidelity is no exception. Natural disposition is aided by environmental persuasion.

Exposure to a certain kind of environment, upbringing, or specific experiences will reinforce or weaken genetic seeds. That's why not everyone will be prone to or actually commit adultery with the same ease or frequency. The mere fact that infidelity is forbidden or frowned upon in different contexts will distort its occurrence.

Those who have experienced parental infidelity within the family are more likely to commit infidelity themselves. A study found that knowing of one's parents' infidelities increases the likelihood of cheating by twofold. A similar effect is observed among children whose parents are no longer married. Having grown up with parents in low-satisfaction and high-conflict relationships also has repercussions. Infidelity has also been linked to certain personality traits, such as neuroticism and narcissism.

When it comes to the genetic counterpart, the heritability of infidelity is far from conclusive. Studies on twins, which aim to separate the effects of genes from those of the environment on a trait, based on the exact genetic sameness of siblings, suggest that the overall genetic contribution to variation in extra-pair mating is 62 percent in men and 40 percent in women. There have been efforts to pinpoint, map, and describe the individual genes involved. But, again, the idea that one gene variant or another might say with absolute certainty whether one is going to be promiscuous is a misconception. What scientists look for is differences, among people, in genes that code for a set of traits, attitudes, or tendencies that pertain or contribute to incidents of promiscuity and infidelity.

One such case is the gene linked to the DRD4 receptor, a brain receptor involved in the metabolism of the neurotransmitter dopamine. Dopamine is a powerful motivator. It also has to do with openness to new experiences and makes us look forward to future pleasures. When we cheat, we take risks. Cheating is also a thirst for novelty. A person carrying an exceptionally potent form of the receptor that facilitates the action of dopamine would be more likely to seek novel experiences and sensations, including the risk and reward associated with having sex with a new partner. One study genetically screened a group of people for differences in their genetic sequence coding for the dopamine receptor and compared them to incidences of promiscuity, conceptualized as involvement in one-night stands, and to broader sexual infidelity. With no differences between men and women, those with the more potent form of the dopamine receptor were twice as likely to have had one-night stands and to have regularly cheated on their partners. In keeping with the role of dopamine in seeking novelty, individuals carrying the same form of the receptor also reported interest in a broader range of sexual behavior.

Men carrying a form of the arginine vasopressin receptor gene that results in fewer receptors in their brains have been found twice as likely to stay unmarried or to experience more relationship crises, with a bigger risk of divorce. A behavioral study revealed the role of oxytocin in strengthening behavior associated with fidelity. When encountering an attractive single woman, married men who received an intranasal injection of oxytocin kept a greater distance from her than did nonmarried single men.

In addition to upbringing, family experiences, personality traits, and genetic dispositions, we need to consider that broader changes in sexual mores influence sexual behavior. With differences at the level of national and cultural contexts, society has certainly become more permissive about sex. "Flings," "hookups," "fuck buddies," "open relationships," and "friends with benefits" are some of the relationship categories that have gained tolerance.

In any case, whether infidelity is encouraged or frowned upon, cheaters have strategies to come to terms with their actions. A common one is to keep them entirely secret, but when exposed there are other, subtler ways. When it comes to memories of problematic moral actions in general—things we feel guilty or ashamed about—it is common to develop biases in recollecting them. We relegate memories of bad actions to the remote past, while we paint a more recent or current version of ourselves in the best possible light. Such a creative way of shaping our biography and self-image is vivid when we are confronted with our infidelity. This has to do with a common phenomenon termed "cognitive dissonance," a sore disharmony between actions and beliefs. We may be aware that sugar is fattening and detrimental to health, so we downplay the fact of having eaten cake by saying that we just eat it once a week and only if it is baked by our grandmother, who uses strictly genuine ingredients. We find ways to deal with our contradictions. A study has provided evidence that when somebody alludes to our being unfaithful, we awkwardly tend to emphasize the rift between our typical self and the behavior highlighted in the insinuation. We also tend to trivialize our actions, as a way to reduce the discomfort that arises from having committed them, as well as the discomfort that arises from our contradictions.

In his delightful book *Conditions of Love*, the philosopher John Armstrong acknowledges a perennial puzzle in our intimate lives. He says that we often chase an ideal of scrupulously espousing deep love with sexual gratification, as if by finding true love in one person we could forever consign in him or her the full satisfaction of our sexual desires. This is not impossible, and when it happens, it is a profound sentiment, a loud expression of devotion. However, as we have seen, sexual drive won't give up its own motives. Armstrong offers a succinct and poignant settlement for this thorny issue facing human nature. He says there are two ways to get around it. One is renunciation; the other is secrecy. With the former, we insist on distancing ourselves from practices of extramarital sex in the name of love. With the latter, while nothing is lost at the expense of sex, love is deprived of transparency. In his attempt to resolve the issue, Armstrong resorts to the notion of dignity and to a story about a duke and a greengrocer from the critic and social thinker John Ruskin (1819–1900). In Ruskin's Victorian times, people unjustly associated being a grocer with leading a somewhat undignified life. That was, of course, a gross and harsh exaggeration. The world was full of grocers who were honest and principled, and who deserved admiration for their work. Ruskin suggested that if a duke sold groceries for a day or two, unfounded assumptions about the lack of dignity of grocers would be overturned. Since a duke's dignity was undisputed, his new guise would help spread the message that it's perfectly possible to be a shopkeeper and be regarded as dignified. Think in these terms about love and sex, suggests Armstrong. Couples struggle because sometimes what one or the other wants to do with sex is deemed incompatible with love. What the ingenious parallel to the two characters suggests is that sex and love, like the duke and the grocer, are both equally worthy of respect and, by turning the argument on its head, questionable in their conduct. We need perhaps a bit more imagination and tolerance in how we see sex and love, or fidelity and adultery, unfold in our relationships. Today most cheaters go for secrecy in the hopes that they won't be caught..

*   *   *

When talking about affairs and infidelity, it's common to refer to love triangles, though, of course, other geometric arrangements are possible. Me, my wife, and the neighbor. My boyfriend, his ex, and myself. My girlfriend, her business partner, and I. It's all about people who, by chasing their passions and each other, expand and narrow angles of contact and rapprochement.

Imagine this in space. Imagine three cosmic bodies that at some point come temporarily into alignment. The intrusion can be blatantly obscuring, as in the case of a lunar or solar eclipse. Other times, the concealment is negligible.

A fundamental problem in the history of astronomy was the measurement of distances. Up until the end of the eighteenth century, thanks to millennia of calculations that culminated with the work by the German mathematician Kepler, astronomers across the world were familiar with the relative distances among the planets in the solar system. They knew, for instance, that Mercury is about four times closer to the sun than Mars, and that Mars is around twice as far as Earth from the sun. But the absolute distance of each planet from the sun, in miles, was not known. They just didn't know how to measure it.

Things changed thanks to a brilliant idea by the British astronomer Sir Edmond Halley, who believed that one very rare event would help solve the riddle: the transit of the planet Venus between the Earth and the sun. Occurring with an interval of about one hundred years, transits of Venus take place in pairs. These paired transits are separated by eight years. Incited by Halley's visionary ideas, the world's brightest minds in astronomy, physics, and mathematics mobilized to observe and measure these rare events in 1761 and 1769.

The transit of Venus allows the measurement of our distance from the sun thanks to a phenomenon called parallax. From the Greek word *parallaxis*, meaning "alteration," parallax provides a way to measure one thing in order to obtain the measurement of something else. When we contemplate an object in front of a background, its position seems to move if we change our point of view. Say we are photographing from afar someone standing in front of the iconic Hollywood sign on the Santa Monica mountains. Depending on where we snap the photo, that person will seem to have shifted position across, as well as up or down, the letters, at one extreme appearing in front of the letter *H*, and at the other in front of the letter *D*, even though they haven't actually moved.

Zoom out to the universe.

If we watch Venus pass in front of the sun from two widely

separated locations on Earth, we will observe a similar effect on the sun's disk while Venus is treading its orbit. In its transit from one edge of the sun to the other, Venus will mark two separate tracings. Speed equals distance over time. Halley noted that by measuring the exact timing of the transit of Venus from different sites on Earth, the distance between those two tracings could be measured with sufficient accuracy. A method called triangulation would then reveal the actual distance of Venus from Earth. The exact calculations are cumbersome, but suffice it to say that in trigonometry it's possible to derive the lengths of all sides of a triangle by knowing the exact length of only one of its sides, if we can also measure the angles to the other distant point. If that distant point is Venus, and we know the distance from the two points of observation on Earth as well as the distance between the two tracings of Venus on the sun's disk, we can derive the position of all the corners of a triangle, hence the distance between them.

The passage of a third body, then, is an excellent excuse to survey the distance between the other two. Cosmic arrangements are such that the alignment of celestial bodies is predictable. Planets just follow their trajectories. In life, an affair is not foreseeable, even though its possibility is lurking. What makes it happen are often accelerations and decelerations in our trajectories, our impulses and hesitations.

*   *   *

The hotel room is small. There's still light outside, and they have ordered salmon and wine. He moves inside her twice and says, "I've missed you," in between. Vanessa says it at the end. The exchange was overdue.

In her book *Mating in Captivity*, the author and psychotherapist Esther Perel, who has advised a vast number of couples on their relationships, says: "Trouble looms when monogamy is no longer a free expression of loyalty but a form of enforced compliance." She notes that, especially in today's society, in which a sort of consumerism of partners is rampant, it's easy to be someone's second choice. Fear and awareness about abandonment can exacerbate possessiveness and a longing for exclusivity. But affairs are a real possibility. It is especially when we repudiate this possibility, she writes, that members of the couple set off to find and play with it on their own, and in endless permutations. So, then, to cheat or not to cheat? And in either case, cui bono?

Without endorsing or justifying infidelity, Perel asserts that acknowledging it sheds light on an important truth about relationships: Each member has his or her own eroticism and sexuality that are separate and independent of their partner's, no matter how committed, stable, and forward-looking their relationship is. By "acknowledging the third," it's possible to demystify its danger and actually bring oxygen into a relationship. While in some cases adultery has devastating effects on couples, in others it does good to the health of a relationship. Some affairs never materialize. Others resolve in one night. A few are impatient and cause devastation. A few others perish and then resuscitate. Some are taken to the grave. Whether it's a daily or a monthly satellite orbit or a once in a while transit of a planet, the intrusion of the third informs us about the real distances between ourselves and our core lovers. Even when they are revealed, affairs do not always imply incurable hurt.

Their respective kids now aged almost seven, more than once Ryan and Vanessa have faced the question of what their union means to their lives, and more than once they have weighed their options. Ryan is never going to abandon Laura but wants Vanessa for reasons beyond his imagination. Vanessa feels the same about Kevin. Their existences are so comfortable in their trajectories, the thought of a perturbation would discourage a titan. Vanessa knows firsthand the complications of divorce. By keeping her infidelity to herself, she is being faithful to her own pain as a child.

At dawn, Ryan takes Vanessa to the train station, on her way to a conference.

"My turn to travel next time."

"Be good, Ryan."

During his short walk home, despite the noise of a street sweeper, he listens to himself repeat those words and the sound of Vanessa saying his name.

When he sees Laura is still asleep, he sighs with relief that he won't have to invent stories about the party just yet. He undresses again, this time to fake sleep.

But there is no need to pretend. Despite the morning light invading his side of the bed, Ryan barely has time to lay his head on the pillow before he dozes. One minute into unconsciousness, he feels a poke on his nose:

"Where have you been, Daddy?"

Split or Steal

(I Give That You Might Give)

When Scott landed on the pier, he was thirsty and had a strong urge to pee.

A man next to him on the ramp asked what brought him to Ireland and how long was he staying. Like a bolt of lightning, Scott replied, "Love."

Scott was a diplomat in his thirties who worked as cultural attaché for the US embassy.

The kind of love that encouraged his journey didn't imply someone special welcoming him on the pier.

For three and a half years in London, Scott had been together with Liam, a biologist his age, whose love for Scott vacillated as soon as circumstances in their relationship demanded commitment on both parts. His mandate in the UK soon to expire, Scott was to be assigned to a new location. Wherever that place would be, he invited Liam to follow him: "Science is without borders. Surely there are labs everywhere . . . like there is always an embassy," Scott would say.

But Liam wouldn't hear of it.

They met in water, an element they both loved. If they wanted to see each other, it was enough to show up at the public pool in the evening, after work, because Liam liked swimming on the edge of the day. He swam fast, faster than Scott. Whenever Scott began to catch up, Liam would pause, on purpose, at one end of the pool and start again when he saw Scott approach, so as to pass each other in the middle. Scott would try to catch him by the leg. Every fifteen or twenty lanes, they would rest and fantasize about swimming together in open waters. Liam bragged about the rebelliousness of the sea in Ireland—"my sea," he used to call it. He liked to mention a place called Forty Foot, the inlet where he had learned how to swim.

"I must take you one day."

"When?"

*   *   *

They never went. In fact, Liam had always avoided showing Ireland to Scott.

When they met, Liam set Scott into motion like a clock. However, from the start the swing of the dial was a question: back and forth, yes or no, near or far, maybe, maybe not. It was like getting ahold of Liam and then being left wanting more. These two men stood on opposite ends of a chasm. Scott needed facts. Liam loathed promises. Scott looked for guarantees, and Liam kept his options open. Scott wanted security. Liam wanted uncertainty.

Scott was fascinated by Liam's experiments, or what he understood of them. Every morning Liam killed a rat, made himself a cup of coffee, then trained other rats to interpret warnings of danger. He made it sound like he was a commander in chief, with the animals obeying his orders. They feared and then learned not to fear. Scott kind of obeyed Liam too. "You are king," he used to tell him.

Liam liked to let on that he was impervious to fear. He was skilled at never finding himself in a threatening situation or in circumstances in which his insecurities would be exposed. At dinner parties with Scott and his diplomat colleagues, he knew how to divert the conversation toward science so only he would be the real expert. Liam did not take criticism well but was quick at finding imperfections in others, especially in Scott. If Scott improvised one of his curries, Liam would lament a missing spice. The restaurants Scott picked were often not to Liam's liking. The music Scott put on during breakfast was either too soft or too loud. When Scott suggested a bike ride in the country, Liam would instead suggest a day at the pool, because biking wasn't his forte. Similarly, Scott's successes and accomplishments were acknowledged but quickly forgotten. Liam, on the contrary, was very sensitive to being overlooked. If he didn't receive the regard he wanted, he would arch into a posture of disappointment, as if his whole body was saying: How can you possibly not pay me attention?

Liam sometimes ignored Scott's phone calls on purpose. When he wanted to hear from Scott, he would occasionally pretend to misdial, so Scott would get in touch upon seeing the missed call. It took Liam a year before he conceded to their being referred to as "together." Scott consistently reserved two tickets for concerts and always waited until Liam could go with him to see the newest exhibition in town. Liam would see shows without telling Scott or would hold off Scott's invitations until productions and exhibitions closed.

Liam conceded to holidays together but would insist on an annual trip by himself, because, he said, he needed the solitude.

Sex between them was irregular and had few or no preliminaries. Liam rushed to the core performance and wanted to get it done quickly, as if making love was a tunnel he needed to traverse and get out of rapidly.

In general, there was an edge in the way Liam opened up to others, an uphill slope you knew hid something on the other side but were not sure what it was, because he first invited you to climb up, then always sent you back down. Liam wasn't good at talking about their miscomprehensions. He avoided confrontation.

Sure, Scott regularly got tired of putting up with Liam's erratic behavior and occasionally protested. But the problem was that being around Liam had conditioned him to comply with his tyrannical whims. Also, the moment Scott showed his anger and distanced himself from Liam, the degrees of vulnerability reversed. Liam would become the most docile creature. The perfect gift, a surprise, a night of sex, and the exceptional cuddle. These short, well-timed sops were often enough for Scott, who would quickly come to compromise. They were triggers that worked like a key on a clock's mechanism, and kept Scott going. Back and forth. Yes or no. Near or far. Maybe, maybe not. Whenever Liam let go for a while, it was only to gain a better hold.

Liam set the terms and conditions of their relationship, and Scott let him do it. Neither could resist their connection, but they had opposite ways of dealing with it. Scott faced it headfirst and begged for every tiny acknowledgement of it. The more Liam felt closeness between them, the more he fled it. Scott craved confirmations of affection. Liam needed Scott too, but always at due distance.

* * *

In relationships, distances are constantly redrawn, especially when there is a tacit or explicit lack of commitment. Intimacy is risk. It's inherently filled with both opportunities and menaces. Emotional proximity bears the potential of profit and benefits but also implies exposure to hurt or disappointment. As we experience intimacy, we negotiate a wish for closeness with anxiety about our vulnerability. One of the greatest efforts in a relationship, then, is to establish and maintain a delicate balance between independence and companionship. Freedom contends with responsibility, need with autonomy.

It's like a duel, a bet, a dicey gamble in which two people calculate gains and losses. Think of countries at war, a chess game, a salary negotiation, or bargaining at a souk.

These are all situations that involve intuition and strategic choices. From the moment we first yield to somebody to when we live with them under the same roof, being in a relationship poses dilemmas: Give or take? Us or them? Free or bound? Alone or together? Share or steal?

Even through fierce and hostile conflicts, rivals know that for either of them to advance their interests, it's sometimes better to cooperate rather than challenge.

Do ut des, said the Romans, "I give that you might give." An agreement whereby something is offered so something may be obtained in return.

However, a selfish competitor may be inclined to sweep the board rather than reciprocate.

For about two years British television aired a daytime game show called *Golden Balls*, at the end of which two contestants competed for a jackpot. This phase of the game was called Split or Steal? as these were the options available to each player, to try to gain the windfall at stake. Their choices, however, remained secret, concealed inside shiny golden balls. If both contestants chose the option of splitting, they would share the jackpot. If they both decided to steal, they would both walk away empty-handed. But if one contestant chose to split and the opponent to steal, only the stealer would keep the jackpot and the splitter would lose everything. Until the players' intentions were revealed in the final disclosing moment, the players had a chance to negotiate. They pleaded or demanded, denied and conceded, persuaded or discouraged. They could declare or hide their true intentions. Or they could purposefully remain ambiguous to lead their opponent on.

So trust alternates with suspicion. Greed wrangles with generosity. Self-interest is measured against a scale of altruism.

Scott had no doubt he wanted to be with Liam. This offer, in all its boundlessness, felt perilous to Liam, who, despite caring for Scott, defected.

Scott shared. Liam *stole*. Their last conversation was over the phone:

"I am sorry, Scott. I believe we do love each other, but I don't feel ready to give everything up," said Liam.

"How can you give *us* up so easily?"

* * *

Liam's defiance, which had always attracted Scott, became the force dividing them.

When two individuals start a relationship, they incur debt to each other. For Scott and Liam the jackpot was their union, built together day after day, choice upon choice. They could keep it or let it go. Any attempt at making it work was also a test of their capacity to at once care for each other and pursue their own personal development.

At stake was not only their union but also their self-worth, a currency which guides the intimacy trade and contributes to orienting the angles and distances in a relationship. Despite redressing itself over time, self-worth is minted early in life.

According to attachment theory, developed by the British psychologist John Bowlby, adult styles of affection and bonding echo one's childhood experiences of connection with caregivers. Whether as children we were looked after or neglected, sheltered or forgotten, pampered or forsaken, counts for our subsequent affective development. Extended to attachment in adult romantic circumstances, the theory goes roughly as follows. People who grew up to available and responsive caregivers are more likely to develop a secure attachment style. Secure individuals grow up confident that they are worthy of love and attention and that when in need of support and proximity, they can confidently rely on others. They are also comfortable with mutual dependence, thinking it is normal to depend on others or have others depend on them.

On the contrary, people who have grown up to caregivers whose availability and responsiveness were inconsistent and unreliable are more likely to be anxious about relationships. Individuals with an anxious attachment style often doubt their sense of worth, don't find themselves lovable, and fear abandonment and rejection. As a consequence, they are clingy. They often seek validation and reassurance from their partners.

There is a third category of attachment style. In the case in which caregivers are consistently absent, unobtainable, and indifferent, children are likely to become avoidant. They grow up to mistrust others and feel wary of intimacy. When relationships pose risks, the avoidant tend to defensively flee and stubbornly rely on themselves. For them, attachment is equated with disappointment so, despite striving for proximity like everyone else, the avoidant are skillful at diverting from it. They deny their vulnerability or dependence on other people because they shun pain.

Attachment styles are sculpted by several competing hands. Early childhood experience carves into the original architecture laid out by an individual's genetic makeup. As we grow up, the encounters we make continue the chiseling. Experience is etched on our bodies, down to our genes.

How life gets under the skin is a phenomenon called "epigenetics."

It literally means "upon genetics." It is a way of controlling how genes translate into traits and behavior that is independent of genetic sequence. It's about how the environment adorns the genes, as it were. Studies with rodents have shown how early parental care leaves a physical mark on DNA.

In nature, strains of mice exist that display distinct qualities of maternal care. Some mouse mothers are just less warm and less dedicated to their offspring than others. Whether or not pups carry genes that make them susceptible to stress or anxiety, if born to a warm and caring mother they grow up to be less susceptible to stress and less anxious. This is because the good maternal care chemically modifies a relevant DNA sequence. One such epigenetic modification is called methylation. Even though this may sound suggestive of an irreversible execution, methylation consists in simply tagging DNA with a methyl group—a carbon and three hydrogen atoms, CH_3.

*   *   *

Scott was anxious. Liam was avoidant. Scott was prone to compromise. Liam was more selfish. At the cost of belittling his self-worth, Scott would always seek companionship. Liam, whose sense of self-worth seemed incorruptible, regularly ignored Scott's needs and preferred freedom. Scott believed in the comfort of intimacy. For Liam, intimacy was a bundle of chains.

Ironically, a powerful magnetism exists between anxious and avoidant people. Though at opposite ends of the attachment scale, the two types feed each other's needs.

Deep inside, the anxious are never certain that someone may indeed love them. They believe there will always be a sizable margin between the intimacy they crave and what a partner is actually capable of providing. An avoidant confirms these beliefs.

When the avoidant are absent and inattentive, the anxious are worried about the relationship. When the avoidant dispense a sudden hint of attention or a loving gesture, no matter how small, the anxious are sent on an exhilarating trip. The temporary joy is falsely interpreted as enduring love; hopes are restored and the anxious are reassured that their avoidant partner cares for them. The intermittent attention of the avoidant, possibly reminiscent of inconsistent attachment figures, is bizarrely nurturing for the anxious. On the other hand, the avoidant match well to anxious people because the anxious reinforce their beliefs that intimacy is an inescapable trap. The desire for independence and the dominating power of the avoidant are legitimized by the neediness and sense of inadequacy of the anxious. The avoidant impose borders so as to trigger a reaction of protest. They go to great lengths to avoid intimacy. Sometimes by making themselves unpleasant.

Unfailingly, Scott was energized by Liam's sporadic signs of attachment. He was addicted to them. Liam, on the other hand, exploited Scott's clinginess for his own self-empowerment.

*   *   *

To Scott, Liam's decision to break up was the announcement of a fate foretold. First he protested and tried to sway Liam. They fought, went silent for a while, and fought again. Then came Liam's nonnegotiable refusal.

Scott succumbed to sadness. He was bitten by regret and devoured by disbelief.

For weeks, he sought an explanation. He insisted on finding words, and excuses, for what happened. Who was right? Who was wrong? He pondered the origins of his persistent leaps of expansiveness, as well as Liam's repeated gestures of dismissal. He found himself doing a sort of accounting of their sentiments, revisiting their

emotional transactions on a ledger, annotating injuries and conces-
sions, blows received and those perpetrated.

During their relationship, Scott had known no moderation and
had always gone for extremes—he either gave too much or nothing at
all. For Scott, love was a higher call, a bid more precious than any form
of calculation. This conviction elevated him. It made him generous, ac-
commodating, oblivious to hurt. When he didn't get enough in return,
he wished he could be like Liam. He ridiculed his own compliance and
wished he could trade it for some of Liam's irreverence. He wished he
could learn how to be stingy with his feelings.

Even when apart for a while, Liam's hold was tight and Scott continued to swing from optimism to resignation about their relationship. Back and forth, yes or no, near or far, maybe, maybe not. Their
union seemed at once plausible and yet impossible. Sure, Scott felt
bitter about Liam, but he also secretly desired that he would show up
again, out of the blue. He longed for Liam's skin and sex. He secretly
hoped Liam would change. Then hopes would inevitably shatter, and
Scott would sigh back into his new condition of loneliness.

There's a poem by W. H. Auden called "The Lesson," which en-
capsulates a constant swing from triumph to loss and declares the
sometimes cruel impossibility of love. The voice in the poem recalls
three dreams. In each of the dreams, two lovers first seem to come
close, but then one hostile circumstance or the other tears down the
edifice of their union. In the first, they are forced out from a house
where they had found refuge from war. In the second, after a kiss, a
strong wind fetches one of the lovers away. In the third, at a tourna-
ment victory ball in which the two are made winners with golden
crowns, they can't join the celebrations because the crowns they wear
make them too heavy to dance. Their union is always impeded and
put into question. At the end, awoken, the narrator faces the lesson

all three dreams maybe suggest: we can't always get what we desire, or love is perhaps a delusion.

* * *

By some irony of fate, about three months after the breakup, Scott found out that his new location was going to be Dublin.

He didn't tell Liam and hoped he would never find out.

The turn of events slowly helped Scott meld sadness into resentment. Liam might have been king, but the monarchy was about to be toppled, he thought. He was ready to end his subjection and see Liam's reign eclipsed. Through his anger, Scott strived to expunge Liam from his consciousness. He wanted to forget. Rather, he wanted to remember better, like Liam's rats. He turned memories inside out. He wanted to catch sight of their most obscure corners, those that like to hide, those he had never wanted to face before. He admitted that, ignoring all warnings, he had developed a stubborn obstinacy in pursuing someone who, in truth, never offered him the tiniest fraction of what he needed. Liam's ambiguity, a quality that in the past had intrigued Scott and later became the leitmotif of their relationship, now made him sick and tired. Scott was keen to regain himself.

Maybe loving and hating someone at the same time is possible. We hate them because they don't give us what we need and what we expect of them.

For all Scott knew, Liam might have hated him too, because he'd expected Scott to run back to him and insist on them getting together, to kindle his ego. In fact, in London, Liam pretended not to be affected by the separation but grinded his teeth, refusing to accept that Scott had not come back to him.

But this time Scott was not willing to accommodate Liam's whims.

On the contrary, he began to regard his move to Ireland as an emancipation. In the past, whatever decision Scott made, he made it for or because of Liam. He wanted to share it with him, hear Liam's enthusiasm about it. Time had come for him to unchain himself and shake off the fear of being rejected, sure there will be other people worthy of his love along his new path.

A study examining the relationship between attachment styles and reaction to breakups has revealed interesting differences between the anxious and the avoidant.

Generally, despite experiencing more acute phases of emotional pain, the anxious, thanks to fruitful reflection, can turn their breakup experiences into occasions for higher personal development. On the contrary, the avoidant may experience quick emotional healing but, in the absence of self-reflection and by lingering in defensive positive self-appraisals, lose the chance to revise their mistakes and derive any constructive meaning from breakups. They remain stuck in their pattern.

Attachment styles are not immutable. They can develop and alter during the course of our lives. Anxious and avoidant individuals can aspire to more secure prospects—alone; between them; or with higher likelihood of success, with someone with a secure attachment style. Equally, a person secure from the attachment point of view might falter and adopt features of either an anxious or an avoidant style.

We can't square a circle. We can't radically, or from one day to the other, reverse deep-rooted habits. However, we can identify items in our or our partner's behavior that beg for change and find ways to work on them. We are far from knowing what epigenetic modifications would need to occur to trigger such changes. Nor do we know how long such modifications would take. However, awareness about the need for change is a good start.

Scott found himself in a special place, amenable to transformation. Significant life transitions offer the opportunity to vividly experience a personalized course of time and to embrace the possibility of erasure and composition. In this rare dimension, we raze and mint several versions of ourselves.

*   *   *

Scott scanned and absorbed the novelty all around the pier. The sound of people talking was the most familiar element. He could detect the song of Liam's speech.

Late in the evening on the same day of his arrival in Dublin, Scott yielded to curiosity and went over to the famous Forty Foot. The inlet was deserted and felt like an alcove. He slowly undressed and edged onto a rock, feeling the salty winds softly rub his skin. Candidly, he sighed. He was at Forty Foot because and in spite of Liam, but also for his own sake, he believed, and on his own terms.

Without hesitation, he plunged, the dark shawl of the sea embracing him, the chills of that instant disquieting yet invigorating. There's something uncanny about the act of diving. It's a short-lived sensation of triumph, a bet won, a promise fulfilled. Self-confidence is put to test. Prospects are condensed in one act of audacity.

According to the principle of Archimedes, the ancient mathematician from Syracuse, when an object pierces a fluid, it is buoyed up by a force equal to the weight of the fluid it displaces. It's as if the object loses some of its weight to lend force to the upthrust. For every risk taken, there are losses and returns. Often, we lose something in order to gain something greater. Scott's resolute move into an unknown future bore all the strength of his pledge to a new life. His plunge carried all the weight of his renewed worth. Whatever he might have lost along the way, he had stirred enough waters to procure change. He had gained more clarity about who he was and the

path ahead of him. Whatever loss he might incur, he felt like a winner and felt comfortable with the victory. He had stolen his own
worth away from Liam's control.

He rose back to the surface, took three or four deep, loud breaths,
spat, and swam out for a hundred yards and back. He then allowed
himself to float silently on his back, all limbs spread out and the eyes
closed. There he waited two minutes, as if to allow his pledge to take
effect, for a spell to be realized:

"You don't exist. We never met. You don't exist. I will forget," he
repeated to himself.

As he opened his eyes again, he felt someone else's presence
around him and fantasized being wooed by a dolphin. He looked
about, and as he turned to reach back for the shoreline, there was
nothing else to do but face the bluff.

Split or Steal?

Still and incorruptible, Liam towered over him from the rocks.
He had changed his mind. He had quit his job and returned home,
to own his territory.

There was no longer a winner or a loser. Only a new bet and
new risks.

Before Scott could do or say anything, Liam jumped in, shouting,
"I told you we would be here together one day."

A Winter Garden

"The truth is boys just don't know what to do with me," said Paul, before biting on one of the salmon cakes.

"What do you mean?" laughed Fred.

"They are attracted, they are tempted to sleep with me, but they ultimately think I belong to a different species, and they don't stick. They run. As fast as they can."

"Is this before or after you have shown yourself in Speedos?"

"Thanks, Rebecca, but usually we don't even get past the first layer of clothes!"

"Come on, pass me the juice and stop being so pessimistic," said Fred.

"But I'm serious. It's like that!"

"Why don't you spend more time online?"

"No!"

"Okay," said Rebecca, "I think your problem is that guys read a sign on your forehead that says Marry Me, and I say this because I might have the same issue!"

"You'll just have to come with us to the club. You'll get laid in no time," said Fred.

"Forget it—I'm not coming along!"

"Hey, and what about me?" said Rebecca. "Do you ever worry if I get laid?"

* * *

Sex was a regular topic during Sunday brunches at Fred's place. How much of it, with whom, and what kind. It was a measure of each person's rank in their circle. It was especially so at the first hint of spring when, glimpsing the prospect of a sunny life outdoors, they felt lonelier than usual, or as Paul put it, suffered from severe attacks of "lust-itis"—that is, they were impatiently horny. All present would trade recipes for how to best conquer their objects of sexual desire and would talk about recent or planned sexual encounters in detailed, even graphic, fashion.

Leaving the table and moving on to the sofa, Fred asked, "What is it that you want, Paul? Do you want a good fuck, or do you want a cuddle?"

"Why can't I have both?"

Unfortunately, for a while Paul had actually been getting neither.

Twenty-nine years old, handsome, and kind, Paul felt he was wasting the best years of his life. Born and raised in Los Angeles, he was at the time he moved to Berlin, for a new job in an architecture studio, emerging from a meaningful relationship that had lasted three years and ended because his boyfriend cheated on him with another man he'd met online. His attempts at finding a new boyfriend, or simply having sex, had been unsuccessful and demoralizing.

Sex begets sex. The lack of it often has the opposite effect. Instead of attracting, it repels.

A successful banker, Fred worked across three continents. He kept a flat in Berlin and returned to it whenever he could because it was in Berlin that he got the most "action," by visiting bars and clubs or logging on to his dating profiles and apps. He basically slept with a different guy every weekend, but he never let any of them stick around for more than one night. Usually, he pushed them out the door the minute he came. Oddly, he then complained about being single.

"Okay, so what about this kid you just met and you like so much—what's his name?" Fred teased Paul.

"Who, Nathan?"

"Right! Do you do crosswords when you are together?"

"Leave Nathan alone. He is actually my biggest and last hope."

"Oh, you fatalist!"

Nathan was a shy young man with a wicked sense of humor and a part-time job as an editor for a travel magazine. Paul had recently met him at a gallery opening. They were kind of dating but having no sex. Fred just couldn't understand why Paul would continue to see Nathan and suggested he drop him. Paul was impatient to know Nathan sexually, but he thought there was nothing wrong with the wait.

After one last cup of coffee, the crowd was ready to test their luck at the disco.

Conveniently, a bus departed from right in front of Fred's doorstep and stopped a stone's throw from the coolest disco in the city, which was open nonstop the entire weekend. Sunday was the new Saturday.

"So, are you coming with us to the club or not?" Fred pressed Paul.

On an impulse, Paul joined them. Ten minutes in, among those who danced, those who drank, and those who made out, Paul was eyed by a stranger who signaled that he wanted to be followed outside, through the emergency exit, onto the lawn at the back of the disco that was a regular spot for cruising. For a good while, they stared at each other from head to toe. The stranger wouldn't yield. Paul's blood boiled, but he was tied.

Finally, with an arching smile that edged on mockery and betrayed embarrassment, Paul stepped forward and mumbled, "What's your name?"

The stranger's reply was unforgiving: "I didn't come here to talk."

* * *

The urge to separate body and mind can prove exceptionally robust. It seems impossible that one could speak for the other without causing misunderstanding.

The theorist and author David Halperin notes that good sex is void of irony. "'Fuck me,'" as voiced in the receding distance between two ardent bodies, "is the least ironic utterance in the world." It demands what its string of letters is there to spell.

Indeed, on certain occasions a smile could only bring confusion. As Halperin reminds us, smiles usually only transpire at the end of intercourse. Before then, they are a superfluous adornment. Affability and blitheness, on which Paul sometimes relied in the attempt to overcome embarrassment, stood as obstacles rather than bridges to the initiation of erotic dialogue.

At moments of highest sexual enjoyment we are typically oblivious to external distractions. We drift toward a seemingly unseizable dimension. This particular trait of the passionate drive can be captured by physiological measurements. For instance, during orgasm the brain's frontal territories, which in general exert control on our emotions, are subdued. The amygdala, which is at work when we are in the throes of fear, also shuts down. The whole experience is raw but obfuscated. Control is lost, worries dissipate. So does irony.

However, no matter how hard we try, mind and body never disjoin completely, and sex, even the most anonymous, never belongs to just one or the other. Regardless of how short, off-the-cuff, or without strings attached a sexual encounter is meant to be, it is loaded with thoughts and assumptions that complicate it. The very fact of seeking sex is aired by hopes and expectations and may be built on excuses, lies, or omissions.

Before we even touch someone, our erotic desire settles on them, or doesn't, in ways that involve not only the most basic instincts and drives that pulse inside our limbic system but also a whole sophisticated framework of cognitive processing in cortical areas. Even at a sex club, in complete anonymity, we respond to a sexually desirable stimulus in a way that involves a certain degree of complexity.

As we roll around, entwined with someone else's body, in a sweat and out of breath, intent on giving and receiving pleasure, a vast amount of information emanates from our bundled past, present, and future. What we touch and what we allow to be touched. The sounds we make, the sounds we don't make. What we say and what we don't say. The turns we take and those we accommodate. The weight of our moves. All this varies according to who we are in that moment, where we are coming from, and what we wish. Our bodies reveal lacks, fears, needs, and trepidations that all our senses, experience, and intuition help us perceive.

Back home, after it got dark, Paul masturbated while watching the neighbors' lights, across the courtyard, go off one by one. The moment after he came, a message from Nathan beeped on his phone:

Are you free to go see a play next weekend?

Instead of falling asleep, Paul stayed up all night.

* * *

Imagine intimacy as a large, sophisticated mansion with a vast number of rooms, large and small, dark and bright, visible and hidden.

For a new guest, the deeper they get into the house, the closer their level of intimacy with the host. We don't let just anyone peek into our kitchen cupboards, much as we don't take them straight to the basement, where we hide our mess and deepest secrets.

A significant part of the mansion's grounds, the garden is a lot of fun. Flowers, maybe fruit, fresh air, and a swimming pool. Good for games of hide-and-seek, it has the advantage of a view and access to other people's gardens. Inside, the house is cozy. A fireplace, sofas, spreads, a library full of books. But there are also drawers full of receipts, bills, and trifles that it takes discipline to discover as well as accept.

Now, both the garden and the indoor areas present risks. Outside, things are unpredictable. Temperatures might suddenly sink, a violent storm might spoil a picnic, distractions may lure us away, and nasty bees may sting. Yet if something goes wrong within the edifice and we are in the garden, we still have a chance to quickly jump over the fence and find an escape. If we are rummaging through old photos in the basement, the whole building may collapse upon us—we may be crushed by the weight of memories. Paradoxically, when it comes to the architecture of relationships, having a roof above our heads is riskier than staying outside. So much for the saying "safe as houses."

Some people's heart clearly resides inside. They have their spot in the basement.

But with today's fashion of endless options, online encounters, one-night stands, and sex available at clubs, a lot of people vastly prefer to keep their guests in the garden. Front door locked and curtains drawn over windows. Entering the house doesn't even cross their minds. However, this means everyone will eventually go back to their own home alone. Safety comes at the cost of loneliness. When we don't share the inside of the mansion, we experience the journey of intimacy only marginally.

Such a strict segregation between desire and attachment is stark in the architecture of relationships. But it is myopic, at best, and a little cowardly in the face of love's amplitude of possibilities. It doesn't do justice to the endless and cunning potential of an encounter between two identities, nor to its intricacy.

Sharing time inside is definitely a less casual involvement than hanging out in the garden but altogether a much more interesting experience that will eventually reveal a lot more about the resident of the mansion.

Culture and upbringing, bolstered by the influence of a good number of religious doctrines, participate in composing a chorus that hopes to persuade us that passion is not synonymous with enduring love. Sex is ephemeral, it whispers, whereas relationships ground. Fucking is different from cuddling. Being friendly destroys sexual tension. From dating etiquette between strangers to sexual schedules in established relationships, the distance between sexual pleasure and romantic affection can open up as a rift. At its worst, the echo of all that propaganda rings like this: We would like to afford to be sexy perverts, but it's tough to be so with those to whom we are attached.

*   *   *

Like many of his friends, Paul just tried to make sense of his intimate life.

On one hand, he despised the universe of online chats and casual encounters. On the other, he was drawn to it, partly because he saw no alternatives.

Most of the guys Paul chatted with online, especially the young ones, were a waste of time. In search of instant gratification, they would entertain long, at times juicy, conversations—strictly by text, because a phone call would be too much to deal with.

The banter was excellent, the rhetoric exquisite—breathless scenarios of erotic abandonment. They sounded as if they couldn't wait a second more to meet him and offered big promises about things they would or would not say or do when they met in person.

But then, after all that virtual excitement and buildup, when it came to arranging an actual date, it all remained fantasy: they either ghosted him or provided excuses that were as extravagant as they were depressing.

Then there were guys who wanted to meet but wanted sex and sex alone.

Once Paul got a message on a dating app from a guy who had once or twice eyed him at the gym.

"Hey, would you like to meet up?"

Though he had never talked to him in person, Paul had heard, from other guys at the gym, that the chap in question was an orchestra conductor who had recently relocated to the city. Paul did fancy him, so a couple of days later he extended an invitation to the theater. Front-row tickets for a show inspired by Shakespeare's sonnets. Dinner and drinks after the play.

Answer: "Thank you for the invitation, mister. It's very kind of you, but I wonder where you are heading with this. Sorry, I am not interested in that kind of meeting."

The conductor ignored the shadow of the shag behind the theater camouflage. The sonnets alone are full of sexual innuendos, but the stranger hastened to withdraw from the game as if, by suggesting a night at the theater, Paul wanted to plunge him into the depths of his life, lock him in there, and promptly offer a marriage proposal. A trip right to the master bedroom of the intimacy mansion, without even a short sniff of the roses in the garden. But a full-fledged date, with theater and the works, must have been too much for someone who, instead of approaching Paul directly, preferred the route of virtual communication. Funnily enough, a week later the conductor asked Paul to have quick sex with him at the gym.

On average, casual encounters based solely on sex are undistinguished. Unless the sexual action itself turns out to be extraordinary, a date that includes a concert, bad wine, a good joke, a disagreement, a flat tire, or—who knows?— maybe even a secret, will leave a better trace than one confined to a Web message and a shag.

Paul was intense. He longed for sex with gusto. He didn't want to simply hook up. He ignored, or refused to accept, the risks connected with venturing deep into the mansion.

Be it for a few hours, a fortnight, or indeed a lifetime, Paul considered himself fully equipped to perform adventurous leaps back and forth from the garden to sundry rooms inside, without making too many distinctions between different kinds of erotic and affective involvement. The result was that any hint of interest from his side was immediately interpreted, as Rebecca had pointed out, as something committal.

It's easy to get the impression that promiscuity is an exclusive habit of the homosexual world. As we saw in "Transit of Venus," this is not so. For instance, the use of online dating for hookups and random sexual encounters is equally common in the heterosexual world. A survey among subscribers of the dating website OkCupid revealed that men and women, gay or straight, have comparable sexual habits. The percentage among site users who were explicitly looking for sex alone was comparable within a range of 6–7 percent, with the exception of a 0.8 percentage for straight women, which, the founder of OkCupid and author of the survey suggests, may speak to taboos around their sexual boldness. Similarities also applied to the reported number of lifetime sex partners. The stereotype of gay men's promiscuity was confirmed only for the outlier in their group. The latter's reported number of sex partners—twenty-five or more—was twice as many as reported for their straight counterparts.

Indeed, Rebecca had encountered similar problems. One of her former boyfriends once sent her a short text:

Sex doesn't seem to work—I am not in love with you.

Rebecca had to read it twice. Then a third time out loud. Their relationship hinged dangerously on a couple of assertions linked by a dash. The terms of the equation left Rebecca perplexed. What came first, she asked herself, the physical attraction or the sentiment? And what counts more, a vast portfolio of sexual positions or an undying talent for caring? Intercourse or a cuddle? She stared at her phone screen, fast-forwarding past scenes of bedtime with him, looking for clues in her and his moves that might have predicted this demoralizing finale.

*   *   *

In the never-ending schooling of our sexuality, which has deep roots, every encounter with a new partner is a lesson. We teach, we learn, we adapt, we forsake. We smooth edges or crease new corners. What we get used to is delicately crucial, for when reinforced, habits recur. Some of us seek sex only; some of us insist on joining sex with love. Some of us are more agile than others at switching preferences.

Though in the past he had enjoyed a small number of random sexual encounters that did not develop into relationships, the larger part of Paul's sexual history was composed of sex united with affection. Not only was this kind of physical experience a habit in the past; it was also a projection into his future. Ultimately, Paul longed for a connection that included but went beyond the sheer physical.

In all its vigor, sex is no doubt beneficial to health and well-being, and is a powerful force. Several studies converge on evidence that the mammalian brain responds well to sexual activity. As a greatly hedonic and rewarding experience, sex is regenerative and cushions the body against the deleterious effects of stress. In the brain, sex contributes to neurogenesis, the formation of fresh neurons. One study in rats revealed that when the animals were freshly exposed to an intense single incident of sex, there was a temporary surge in stress hormones—perhaps related to the novelty of the encounter—but there was in parallel a clear and significant new growth of neurons in the hippo-campus, a region of the limbic system that is involved in emotion, the formation of memories, and spatial navigation. When the exposure to sex was repeated—sometimes with the same partner—neurogenesis was boosted. The sexual experience also reduced the animals' level of anxiety-like behavior, as measured by a series of lab behavioral tests. Opiates and other molecules, such as dopamine, involved in stimulat-ing pleasant reward sensations, as well as the bond-inducing chemical oxytocin, partake in modulating the neurogenesis and anxiety-reducing effects.

Now, sex by itself is good, but it is no less good if enveloped by a sentiment of security, belonging, and understanding. In both the short and long term, emotional intimacy does nothing but improve sexual life. Longitudinal research begins to suggest that communication, as well as commitment and a sense of stability within couples, both mar-ried and unmarried, improves sexual satisfaction, as well as the overall quality of a relationship. One study that specifically explored this link in newly married heterosexual couples found that communication, self-esteem, and relationship stability prior to commitment and nuptials were crucial for ensuring sexual satisfaction in the early stages of mar-riage. A wife's confident ability to communicate her sexual desires helped her husband understand what to do to please her sexually. In turn, a husband's willingness to communicate empathically and under-stand his wife's feelings largely governed her sexual satisfaction.

One of the excuses Fred used to explain why he never met again the boys he slept with was the bad quality of sex: "We didn't click at the level of skin." Performance was a crucial issue: "Suppose you sleep with a guy and it doesn't work," he said. "He's clumsy or, worse, you are clumsy or can't get an erection or reach orgasm. If this happens with someone you don't know or have just met, it doesn't matter. They'll leave and that's that. No shame, no embarrassment."

There can be a considerable distance between expectation and likelihood, between fantasy and the reality of sexual intercourse. With our imagination, we can go very far, but reality is far from perfection. Interestingly, this clash is reflected in electrical activity in the brain. Similar parts of the brain are involved in experience and imagination, but in each case the flow of information between those parts changes its route. When an image reaches our eyes, the visual input is first received and interpreted in the occipital lobe. This is at the bottom of the brain, at the back of our head. From there, the signal moves forward and up toward the parietal lobe. When we imagine something, the flow of the electrical signal is reversed. It skids from the parietal lobe down to the occipital lobe. So actual visual perception and mental imagery flow in opposite directions.

Sex doesn't need to be fabulous the first time we sleep with somebody. Sexual skills and compatibility improve over time, and emotional intimacy has the power to accelerate the process. Better sex can arise from comfortableness with one's partner, and qualities such as openness, patience, and trustworthiness can benefit, and not threaten, sexual satisfaction. Basically, there are greater advantages in enjoying sex, even wild sex, within the comfort, wisdom, and kindness of a loving bond with one's partner. Or at least one can try.

Though he wouldn't admit it, beyond the urgent, purely physical need to be touched and exchange fluids, Fred also needed to be held, cuddled, understood. He had just gotten used to fucking around so much, that he acted as if he wasn't able to sleep twice with the same guy.

* * *

Next to those who only wanted sex, and those who only fantasized about sex, another category of guys, often with a number of traits and qualities that made them perfectly eligible long-term partners, did go on dates with Paul, maybe even slept with him, but then, despite their attraction to him, suddenly withdrew, offering the following sort of lip service:

"You are perfect, but I am not the one for you."

"Oh, you are so interesting—you need someone more special than me."

"You are unique. I have never met someone like you."

"You know, with you it wouldn't be as it is with all the other guys. With you, it would be different."

Oddly, rid of Paul, all of them would rush to find their next partner, with the same hope of keeping them, although often they would turn out to be another mismatch.

Unique, perfect, interesting, different, special.

In truth, all these flattering attributes were a cover for something else: real. By rejecting Paul, these boys were avoiding desire itself. One quality they were still missing was the maturity to have an open heart. Altogether, chats, sexting, ghosting, and one-off dates work to keep everything in the realm of impossibility, to stay safe from the intensity of real desire, and from the anxiety connected to it, which imagination does nothing but exaggerate by fanning expectations.

Some believe that the gay world in particular has had, by necessity, because of the repression it suffered in the past, to originate a culture of intimacies—cruising, parades, promiscuity, fuck buddies, etc.—that has little in common with domesticity, coupledom, or family. It's as if these intimacies were created to counteract a status quo of intimacy to which gay relationships always stood in deferential position. Sadly, it's even possible that some guys, for various kinds of psychological pressure from society, or even family, skirted Paul because deep inside they still weren't at ease with their own homosexuality. They got stuck in the belief that they themselves and what they wished for were not worth it, and that a relationship with another man was undesirable, so it was better to run away when a real prospect loomed.

Behavior is contagious. Unfortunately, these attitudes propagate, generating reticence, cynicism, disbelief, and a kind of forced resignation to the impossibility of love. At the high price of loneliness.

Slowly, Paul realized that by suggesting a real relationship to the wrong people he was himself avoiding intimacy and making himself prone to rejection and disappointment. Paul got fed up with having to compromise. He resolved to aim for the intensity he desired. Everything else would have been a surrogate for what he wanted. He longed for someone who was also willing to take risks, with courage, curiosity, respect, and a sense of adventure.

Going back to the metaphor of the mansion, perhaps all Paul needed was an attractive and spacious winter garden. Invented for recreational purposes, a winter garden is an environment that can hardly cause claustrophobia. Filled with exotic plants and flowers, a winter garden embeds us in the rawness of nature but is at the same time a protected, private space. It's at once cozy and unfamiliar. Halfway between the unbridled follies—real or imagined—of the outside lawn and the specter of stifling commitment of the inside, the winter garden is close to the door but also steps away from the master bedroom. A winter garden is a dimension where we might accept our own and someone else's offers and limits of intensity. A space where two emotional and sexual histories intersect, mixing libidinal talents, fears, and aspirations.

The next time Fred challenged Paul for not giving ample space to his sexuality, Paul replied: "I do like sex. A lot of it. I have no problem getting dirty. I just want to get dirty with my boyfriend!"

* * *

It turned out Nathan was a virgin.

Sex with a man streamed through the circuits of Nathan's imagination, but never with anybody had images turned into facts. In Paul, however, Nathan thought he had found the right person to venture a step.

They went to see a play, then on another date, and several more. They watched movies and ate pizza. They hiked and jogged. They chatted about travel and all the places they still wanted to see. They talked about courage and about manhood. Nathan would ask; Paul would answer. They walked each other home and texted before going to bed and in the morning. They quizzed each other on *New Yorker* cartoons to see which one made them laugh harder. Nathan cooked for Paul, and Paul played guitar for Nathan. For weeks, they talked and had no sex.

At some point, Nathan shared a message. Or at least a message became clear to Paul. While Nathan skirted any direct hint at full sexual intercourse, his moves began to suggest that he was ready and that he desired to be . . . taken. Forthwith, without warning, and with no reference to it.

It took a few tries, but eventually the gist of his wishes became apparent. On several occasions, Nathan stood in front of Paul and kind of waited. He wouldn't say anything but faintly paused for a moment that was long enough to suggest he expected something and conveniently short enough to let on he didn't. Along the river, in front of a painting, before a fire, at the window.

"Our desire," says Freud, "is always in excess of the object's capacity to satisfy it." It's as if we consigned lust to an impossible actualization.

As if we always went for more than we can have or could actually handle. Yet we long. And when we don't get what we fantasize about, we feel frustrated.

One of the most tormenting, but arguably most rewarding, challenges in an intimate relationship is to succeed at interpretation. To act and be in agreement with the other's wishes so that we inch toward the possibility of anticipating them, toward the possibility of feeling what is not obvious to anybody else. To understand intimations, beyond and through.

Going back to Halperin, good sex is void of irony, but love is annoyingly full of it. There is a distance between the experience of love and what lies behind it, discord between its notes and the song we hear of it. Sex is a temporary suspension of this dissonance. But ultimately it's in the exercised knowledge of each other, in the interpretation skills we polish in the spaces around, through, and between sex, as we try to make sense of who we are, alone and together, that we learn how to cope with the noise or at least to reduce it.

In other words, we might say it takes a lot of dexterity with irony, and resignation to its tricks, to be able to let irony occasionally disappear for the sake of good sex.

The Italian writer Italo Calvino wrote that there is actually a "profound connection between sex and laughter." Laughter, he says, is "a defense of our human trepidation in the face of the revelation of sex; it is mimetic exorcism to enable us to master the absolute turmoil that sexual relations can cause." It's a "recognition of the boundary that is about to be crossed, of entry into a space that is different, paradoxical, and 'sacred.'" Also, mysterious.

On the next occasion, as they stared out the window, Nathan didn't speak, and Paul remained silent too. He also didn't smile. If he was the protagonist in the theater of Nathan's fantasies, he needed to live up to the demands of the scenes.

It was up to him to begin.

Their bodies uncharted territories, the two went discovering without a set direction. From innocence to experience, from strangeness to familiarity. They surveyed and they mapped. They climbed.

So, here. Now. Show me. The past. Across. In a moment. Let's go. There. No. Yes. Less. I don't know. Heavy. I see. Good. You said. I promised. Ask. Future. Except. Light. I can't. More. For you. Naughty. Despite. Because of me. Air. Space. Down. Ground. Further. All around and wide.

A Wizard's Farewell

"We're all right—aren't we?"

"Yes, of course we are all right, my love."

"Are you sure?"

"Certain."

Oscar began asking his wife, Margaret, questions like these from his bed when, toward the end of his days, he tried to die.

"Then I shall go. What do you say?"

"Right," she would reply. "Go on!"

"Are we really all right, precious?"

"Fine, my own . . ."

His eyes locked onto hers, looking to find complicity in his final march:

"Let's go, then. Let's do the magic."

As if reciting a formula, he sat up and shouted: "All right, close your eyes, Margaret. One, two, and . . . three!"

With his arms raised to the air, he waited for a few instants until it was clear that nothing had happened. Death had ignored his invocation, and both wizard and apprentice remained exactly where they were. Disappointed, Oscar sighed, slumped back into his spot under the sheets, and with a kind gesture of his hand, begged Margaret to leave him alone.

A captain during World War II, the eighty-five-year-old man was among the US troops who helped liberate Sicily from the fascists. The locals in the town where he landed called him *biunnu pagghiazzu*, "blond clown," because he was easily made to laugh. The military triumph and the empowerment he derived from having helped strangers regain their freedom infused undying optimism and a sense of promise into his veins. He looked forward to raising a family back home and swore to himself they would never succumb to gloom or hardship. Among the speeches, the marching bands, and the dances at the victory parade in Connecticut, he noticed Margaret, one of the girls who had decorated the streets with flowers and flags. They married six months later, after he got a job as an assistant in a law firm and she was hired by a children's charity. Oscar went on to study and later became an impassioned professor of law and political science. Together they had one daughter, Amy, who became the hinge in their lives.

Like enemies gradually invading a foreign land, metastases of a stomach cancer insidiously spread throughout his body. The cancer being inoperable, and its diagnosis late, there was nothing else for Oscar to do but wait for his exit and make his last days as meaningful as possible. A tenacious man who, through hope and sacrifice, had given everyone the impression of being eternal was now facing the end.

When the doctors warned of his imminent decline, Amy left her studio in Los Angeles and moved back in with her parents, in the house where she grew up. She had always been her father's most faithful soldier, and the time had come to defend and stand by her old commander on his most difficult campaign. A woman in her late forties, Amy was a painter. An aura of grace surrounded her. Tall and with long red hair, she always looked fresh, as if she had just emerged from a bath.

"We called you Amy, but we should have called you Venus, because you look like her!" Oscar used to embarrass her by showing her the Botticelli painting of Venus in an art atlas in his study. Oscar adored his little girl and made no secret of it.

She was born prematurely. Oscar was at home when she unexpectedly arrived, early one morning in spring, while Margaret was visiting the hospital on her own for a regular checkup. The old espresso machine he had brought from Sicily was steaming its morning load into the cup, when Amy's sleep in the womb was interrupted. Minutes later, her head peeked out. A few grunts and one long push, Oscar was told, and she quickly emerged. On her first evening, Oscar and Margaret's eyes, and those of her godparents and other relatives, hovered over her cradle.

Now, perspectives reversed, Amy looked over her dad from above. At the end of Oscar's life, it was now Amy who took care of him. When she found him relegated to bed, she had to make a huge effort not to break down in tears in front of him. She cooked for him, she opened and closed the curtains in his room, she fetched flowers, and she put on records.

She had promised him, and herself, she would paint his portrait while she was there—an idea that she had toyed with before, but never had the courage to face, and now the mark of Oscar's disease could not discourage.

The day Amy arrived, Oscar asked her to put on the music that had bonded them through the years: tunes from after the war.

On Oscar's bedside table, in a simple silver frame, stood a black-and-white picture of Amy as a child. He had captured her in the moment of discovering her own shadow, when she was about four. With her head only partly turned behind her, and on tiptoes, she swung her skirt in the air as if to obscure the sight of a monster. Her expression wasn't scared. She looked entertained and annoyed. Her eyes glared as though she wanted to defeat it. As she grew up, every time they looked at that picture Oscar would remind her of the incident.

"You see how combative you were? Remember that shadows will always chase you. Don't let them confuse you, but always follow the path you believe is best for you."

Growing up, Amy indeed learned that shadows multiply and chase us in all directions, and that it's not always easy to dispel them with a laugh. When Oscar wanted to know how his daughter was doing, he never asked her directly. He would say: "Are there any shadows, my child?"

One of these was chasing her now that her dad was about to depart. She had always wanted to please Oscar. Despite all the love, the affection she had always been certain of, she wasn't immune to doubts about Oscar's appreciation of her talent, her career, her position in the world.

Partly because he was loving but did not excel at expressing his emotions, and partly because she believed he would have liked her to follow his footsteps in academia, this kept her preoccupied. Nobody took him seriously, but he sometimes joked that he wished Amy would one day become the first Madame President of the United States. First a soldier of the state and a later a professor, Oscar had always been tied by his career to values of economic and domestic security. Sacrifice had been his lifeline.

Amy feared Oscar overlooked who she had become and concentrated on who he wanted her to be. She hoped she would have a chance to chase away that shadow once and for all, before it was too late.

*   *   *

"Ouch, you are hurting me," said Oscar, as Amy tried to clip his large toenails.

"Sorry, Papa. I'll be gentler."

"Please, sweetheart!"

"I'm almost done. Listen, we've got to do this or they'll grow in!"

When she was done with his nails, she would massage his legs with cream, bend his knees multiple times, and rock him to and fro to aid his blood circulation. Then she would comb his hair. At first, it was awkward for Amy, and for Oscar as well, that she should touch her father's body like this. It was something neither had ever imagined would happen. But although Margaret had suggested that only she should do it, Amy insisted on alternating with her mother. These gestures of care and attention became small, essential gratifications for Oscar, who was very appreciative of them.

Oscar's bedroom became a gallery. He requested objects from the rest of the house be close to him, together with the memories they embodied. Some just helped him keep up his daily habits. Others carried a certain sentimental value. He wanted the large TV so he could watch the news. His favorite sofa cushion landed on his bed. From the kitchen, he got his initialed grandfather's wall clock. On a small table, together with his vinyl collection, stood the old family gramophone.

The progression of Oscar's cancer coincided with the onset of Alzheimer's disease. Both rampant in their intentions, cancer cells kept multiplying, neurons deteriorated.

Like cancer, the course of dementia is difficult to predict. It's an unruly condition, an illness that takes surprising turns. For Oscar, short-term memory declined first. His recollection of mundane events was full of holes. After a meal, he would forget he had eaten and ask again for the same dish. After brushing his teeth, he would ask for his toothbrush. Oscar's thoughts tangled and memories blurred. He also took on a few bizarre habits, like repeating sentences or saying things that made little sense. The cutest of these had to do with Amy. Each time Amy entered the room, he would shout at her: "Eh, you've come a long way, little girl!"

Margaret helped him keep his mind active. Research has shown that familiarity and repertoires of shared experiences make it easier for intimate elderly adult couples to aid each other's memory in the face of cognitive decline. Couples remember details of events better together than when alone or with a stranger. Margaret would play games and crosswords with Oscar, and would ask him to help her write down shopping lists before she went off to buy groceries at the supermarket. Both she and Amy also enjoyed reviving memories of past holidays taken together.

Talking about politics and current affairs also kept Oscar engaged. It seemed he had saved most of his lucidity for that. Their household had always been filled with fascinating personalities from all corners of society, and even during Oscar's illness, Margaret and Amy tried to maintain that atmosphere. They invited friends over for dinner or for afternoon tea, to entertain him. They had lots of friends. They were people one could always count on for advice and support. Their door was always open, especially in cases of emergencies.

At first Oscar didn't want to die. Soon after he was diagnosed with cancer, he did not once explicitly acknowledge his precarious situation and always tried to avoid the topic of death. He acted as if the illness didn't, in fact, concern him. He said he had barely become acquainted with the basic strokes of life and humanity, and was hungry to experience more. He was too curious to know what was going to happen in the world, mostly from a political point of view. Will the Middle East conflict ever end? What would become of the

European Union? What human rights campaign still needed to be fought?

However, when Oscar understood there was really nothing left to do to save him from death, he conceded clues to his awareness of his imminent departure. Little by little, the subject of death crept in and, like when he played being a wizard with Margaret, it surfaced in original and delicate ways.

"Am I not too young to die?" he would ask.

In response, Amy told him that as a commander and a father he unfortunately had to lead the way and go first: "Pave the road, Papa. Mama and I will follow you one day."

* * *

Fathers are singular. Proverbially—though not universally—figures of authority, dads embody and impart discipline. Sometimes legendary, irreproachable, and possibly enigmatic, they often succeed at maintaining an unrepeatable rank.

In a moving essay entitled "My Father/Myself" novelist and writer Siri Hustvedt begins: "There is a distance to fatherhood that isn't part of motherhood."

The incontestable divergence is noticeable in the early days after birth. Across the animal kingdom, tending fathers are a minority. Male sea horses nurse; penguin dads brood. Only 3–5 percent of mammalian fathers actively engage in early parental care. Marmoset fathers raise their offspring. So do tamarins. But in most cases, from mice to rats, dogs to gorillas, fathers' roles in child rearing are rather subsidiary. Although we don't lactate, we human fathers do share in the raising process. However, it is often the case that the attachment between fathers and their children has time and room to ripen more gradually.

Animal studies of parental and infant interactions have concentrated largely on the role of the mother, not least because it is difficult to define and quantify paternal investment. This is because some of it is indirect—such as defense of the nest, procurement of food, as well as forms of teaching that include launching young into the sphere of socialization.

In humans, too, there are differences in the way mothers and fathers interact with children. Some of these were shown in timed-series observations of mothers and fathers who tried to synchronize with the nonverbal language of their infant child. Mothers typically engaged in steady face-to-face exchanges characterized by mutual gazing and shared facial expressions, as well as co-vocalization. Fathers engaged rapidly in more random peaks of play rich in excitement and novelty. In the brain, parental care is roughly mapped onto two main neural networks. An emotional and motivational aspect of the parent-child interaction involves primarily the limbic areas—including the reward center—which constitute evolutionarily older sections of the brain. A more attention-based aspect of parental care, which entails social cognition and empathy, is mapped onto the same areas of the cortex that are at work when we for instance try to read other people's minds or identify with them. In an experiment in which parents were exposed to videos of their own child playing, the brain reactivity differed somewhat between mothers and fathers. Generally speaking, and with a degree of overlap, mothers responded with more pronounced activity in limbic areas, while fathers responded with greater activity in the social-cognition areas in the cortex. This might have to do with the fact that from an evolutionary point of view, in light of infant survival, the role of mothering is more ancient and primordial, and also points to the conventionally subsidiary role of fathering.

Interestingly, the parts of the fathers' brains with heightened activity include areas, such as the inferior frontal gyrus, which are located within Broca's area and involved in language comprehension and production. According to the French psychoanalyst Jacques Lacan, the father initiates the child to language, thereby introducing her to the symbolic order of the world. With variation, this transition begins at around the second year of age, which is when dads become more present in their children's lives. Though this might sound far-fetched, and is only speculation, it is intriguing to think that Lacan's symbolic order may be mapped onto the language areas of the brain and that—whether by culture, spontaneous disposition, or both—these coincide with regions in the brain that are more active in fathers interacting with their children.

Amy was the son Oscar never had. Or at least for a long time that is how he preferred to look at her. It was Oscar who checked on Amy's school assignments and dispensed his praise or recommended she be more precise in her work. In his study, she lost herself among heaps of books and walls covered with paintings. From his hands came the heaviest tomes from the top shelves and stocks of crayons, paper clips, and fountain-pen cartridges he kept in locked drawers. It was with Oscar that Amy first discussed God, politics, elections, and war. It was "Papa!" she screamed, dazed, when she first saw a Rothko and for rescue when she fell off a horse. Papa's eyes were the beacon she'd looked for in the audience when she was handed her college diploma and when she was dressed up and ready to go to the high-school prom with her first boyfriend.

Oscar's face loomed in her mind the first time she sold a painting.

The two had something else in common, a ritual that had bonded them throughout the years. When Amy was a child of ten, she and Oscar took on the habit of dancing together at home. The improvised dance floor would open early in the morning before school. The act was disciplined. First they listened to the news and discussed it at the breakfast table. Then, as a reward, they would get up and practice their steps. The music was from Oscar's youth: wartime tunes that sent him back to the period when he'd flirted with Sicilian girls in his regiment's tents. As they danced, Amy relished listening to stories from those heroic times.

Margaret was not a part of this. It was entirely a father-and-daughter affair.

* * *

A couple of weeks after Amy's arrival, Oscar's condition worsened visibly.

He started to have hallucinations, which had a clock of their own and later turned into frequent delirium. He saw flying soldiers that multiplied. He shouted at them, urged them to land or disappear. He talked of bombs, kidnappings, spaceships.

A common symptom of dementia, delirium alters consciousness, misguiding it as to whether the person is awake or asleep. Delirium is also a disturbance of language. Sometimes Oscar was perfectly articulate. Other times, his speech was so confused it seemed he used a different alphabet. Like its language, delirium's neurophysiology is difficult to decipher. The teetering of consciousness and awareness between the thresholds of wakefulness and sleep is marked by neural projections between the cortex and the brain stem, at the back of the brain. At the crossroads of these projections sits the thalamus—the word is derived from the Greek, and means "bedroom"—which, aptly, given its rather central spot in the anatomy of the brain, integrates sensory information of both emotional and cognitive origin. The thalamus works like a sifting portal. If the filtering system fails to function, an input overload will cause psychotic symptoms. Partaking in this transmission of information is a pool of neurotransmitters, especially one called acetylcholine. Acetylcholine promotes focused awareness and attention by acting as a modulator of the filtering system. A dissipation of cholinergic transmission (i.e., involving acetylcholine) has been linked to the occurrence of delirious symptoms. A reinforcement of dopaminergic activity (i.e., involving dopamine), which dampens cholinergic activity, also leads to delirium. Dementia, Alzheimer's disease, and other conditions of cognitive impairment are characterized by reductions in the functionality of the cholinergic system. In Oscar, the combination of his dementia, morphine, other sedatives, and cancer drugs all contributed to his warped thinking.

Delirium is unpredictable. However, it tends to be nocturnal. One night Oscar was particularly restless. It was summer; the house was warm, and a dry wind blew. At around four o'clock, Amy heard him shout. Margaret slept on undisturbed, and Amy didn't wake her up. She ran downstairs to be with Oscar, and in doing so she was to experience one of the most intimate moments with her dad.

It was difficult to understand what he was saying, but he delivered a frantic, confused rally that sounded like a chronicle of a military action, full of tension and pain. He gave and received orders. He called out danger. He shouted for help. It sounded like he wanted to save people or save himself. *"Non uccidete più,"* he shouted in Italian. *"Non uccidete più."* That is, "Don't kill again."

Then, his speech became one of fear, a fear of being lost:

"Don't leave me here. Don't leave me here. Don't leave me here."

When he realized Amy was next to him, his expression first arched with surprise and then distended with great relief. The rambling battle dissipated and so did his pleading. His speech was muddled and mired with impediments, but without taking his eyes off Amy, he began reciting something from the past: "Love of the art and others . . . fear of death . . . To paint's to breathe and all the darknesses are dared . . ."

As Oscar struggled to speak, Amy held his hand firmly. Her eyes welled up. These were lines scrambled from "Rembrandt's Late Self-Portraits," a poem by the English poet Elizabeth Jennings. It wasn't a random choice. It went all the way back to when Amy was in sixth grade. An assignment required her to choose a poem and recite it in front of her teacher and class. Undecided about what poem to pick, Amy received the suggestion of that particular poem from Oscar. Speaking to her budding passion for painting, the poem fascinated Amy deeply. She learned it by heart quickly and recited it proudly, leaving her class spellbound.

In those unique moments of connection, by means of a shared meaningful memory, Oscar reminded Amy and himself that his end was near and his greatest legacy was his daughter, who, being an artist, would live for a long time.

Oscar chose words that celebrated craft, art, and the legacy they leave behind in the world. Through Oscar's delirium, something conjured up the memory of that particular poem.

In his book *Hallucinations*, Oliver Sacks describes the possibility of delirium being a midwife for "moments of rich emotional truth." It may unexpectedly deliver discoveries and disclosures. On a night of alarm and confusion, all Amy was given was a fragment, a small piece of the past. However, she needed no further word or gesture. Just as those lines had been meaningful when she was a little girl, they were now a seed for the future. A future without her dad but also without doubts about his approval. The next day she rushed to finish his portrait.

* * *

When Oscar was at the point of death, Margaret and Amy faced a dilemma. He suffered, and it would be less painful for him to depart soon. Doctors and friends had told them that sometimes the dying won't leave because the presence of their families, with their grief and their prayers, won't let them. So even if Margaret and Amy cringed at the prospect of not being around during Oscar's final moments, they regularly left the room to leave him alone.

Knowing that we are going to die alters our perception of time. While the days and hours shrink, life expands. It becomes richer. While being engulfed by the shadow of an imminent end, a rare intensity creeps in that is perceived by both the dying person and those who survive them. In the face of death, what we know as life takes on an urgency that is otherwise easily forgotten in routine. An urgency for living authentically knocks at the door of the dying person and those who are going to be left behind. Death humbles and makes certain buried matters more urgent than others.

The coordinates of intimacy shift too. The approach of death resets distances. Often friends and acquaintances withdraw, out of discretion or fear. But the closest people, those who'll stay until the very end, become even closer. They look for meaning. They look to care.

Death and dementia continued to compete for Oscar. By meddling with Oscar's lucidity, dementia scarred and imprisoned him. Death wanted to purify and free him. Even as the cancer destroyed Oscar, even as dementia gradually shrunk his mind, a spark of his individuality remained, and so did his need for proximity and affection. Indeed, in the dying the desire for acceptance and recognition, their yearning for love, is almost childlike. They expect unconditional love. But even the smaller gestures count as the greatest, because needs resize too.

The dying tend to return to their purest. They shed superfluous adornments, sometimes attributes that were given to them by other people.

Amy was convinced that even when he hallucinated, a residue of

Oscar's identity remained intact, that he hallucinated his own way. Only so could he have given her the wonderful gift of the poem, which sanctioned their mutual respect and worked as the most loving of good-byes. That's what Amy learned through sharing her father's last days—that life is not about being brilliant, successful, or powerful. It's about owning a spark of individuality. It's about recognizing and being oneself, always.

Oscar became quieter and quieter. He was tired and felt defeated. After the night of the poem, there was more delirium, but there was mostly silence. A silence that bore a message. A loud voice emanated from his subdued glance. It seemed to say:

Take care of each other.

When I'm gone, pretend I'm still here.

Be united.

Don't ever waste your time in arguments.

Keep the house full of people.

Continue to laugh.

Keep creating.

At the moment Oscar ceased to live, Margaret and Amy were gathered around him.

The music was on. Margaret held his hand. He gasped and rattled, but his heavy, tired breath thinned and became quieter.

"What do you need, Papa?"

Oscar raised his eyes toward Amy and Margaret. Amy embraced him and he just slurred, "This."

Then he went for it again:

"Shall I go?"

"Go on, my own," said Margaret.

And this time the magic worked.

Equal

Lev heads home after the Wednesday afternoon lecture.

It's fervent times and a budding week in May, the entire nation a promise. On Friday a referendum will take place, which if passed, will make same-sex marriage in Ireland law and for the first time in the world, so by popular vote. Students call their grannies to make sure they'll vote yes. Straight footballers hope to be best man at their gay team members' weddings. Parents defend their unconditional love for their straight and their gay children. The elderly want everyone in future generations to have the same rights they had. Children wish they could vote to help. Canvassers spread messages of solidarity door-to-door. On the streets, campaigners hold posters and give out badges: Tá–Yes.

Lev's short walk is punctuated with chats and a quick drink with friends. He knows so many people that there is always something to do. Together, they are all enthralled in a ripple of courage and change. Even if the vote's outcome is still uncertain, their unity and belief make Lev feel he can trust an entire nation and that whatever happens, he won't be alone. He feels strong and ponders what an idiot he was every time he gave up hope in his life.

It's during moments like this he wishes Fionn could see him.

At home, Lev takes his shoes off without untying the laces and runs barefoot to collect the washing from the line before the purple clouds dissolve into rain. Plato barks after him. On the way to the kitchen, he gives one last look at a stack of books he has finally decided to get rid of. He dices the onions, melts the butter, and opens a bottle of white wine. While the risotto soaks in the broth, he turns the radio on and plunges into the striped velvet armchair, from which he gets a glimpse of the city bay. Plato curls at his feet. He sips wine and waits for the news.

He can't believe his ears, but there is no doubt. From the microphone, a voice from the past, which now expands through the city air, enters his bones.

Immediately, flashbacks in time and space rush to him, shutting his eyes.

That same voice had once delivered a message that sounded unforgiving and yet became the greatest motivating force for Lev.

Shakily, three years before, that voice had said: "Maybe certain things will never change."

*   *   *

The first date was a concert. A composer renowned for being better live than on record. Fionn made the suggestion at the gathering marking the end of the winter semester.

"I believe the music would enthrall you," he said, making sure those around him noticed he was bold enough to ask his former teacher out, now that classes were over.

Without blinking, Lev said yes.

In the darkness of the concert hall, rather than on the music Lev concentrated on feeling Fionn's presence and hoped he wouldn't notice. His breath, every small movement. What would it mean if Fionn stretched his leg, tapped his fingers, if he closed his eyes and for how

long, if he swallowed or sighed, and was his leg going to touch Lev's? It did. Inquisitively. Then Lev put his arm around Fionn. Neither of them took part in the final applause. If a body exerts a force on another, the latter exerts a force of equal strength, says Newton. Action and reaction, in opposite directions, and nothing between them. Later, at a bar, time flew. Neither of them realized it was four o'clock in the morning until the waitress brought the bill and told them they could have one last drink, on the house, she said, because together they "looked right." Before leaving, they took a leak. Lev entered the toilet specifying loudly that he didn't know how to piss in the presence of others, but he leaked all right when Fionn, already at the urinal, said he was in fact having the best piss of the day.

Out on the streets, the air was a razor. Neither knew how to voice his desire not to part, nor did they know how to stay together. Lev considered saying "thanks for the lovely evening," but held that sentence back, thinking it would be obvious, maybe too polite. Fionn just blushed. He expected a move of some kind to come from Lev, who didn't deliver and was immediately stung by regret.

The following afternoon, Lev called Fionn and invited him for a walk.

When Lev reached Fionn's apartment, to pick him up, the music from the concert was playing. Half a knock and the door opened. Reading glasses made Fionn's eyes look huge, Lev thought. Fionn noticed Lev had ironed his shirt. The desk was crammed with open books, piled high, spines up. There were clothes sprawled on the bed, plants and flowers on the windowsill. A fresh scent of foam emanated from the bathroom. They left the music on.

On the stairs, on the sidewalk, and the dirt path, Fionn covered moments of silence by humming a tune. The park was not crowded, the sun about to sink among the treetops. At the lake, they only walked half a circle, Lev in a coat, Fionn wearing a T-shirt.

"Here, pose. I would love to have a picture of you."

Whatever desire or question belied it, it was an altogether brave invitation on the part of Lev, who had little to no talent as a photographer. There was a bench by the shore, and Lev instructed Fionn to sit on it, the reflection of the water would be perfect for the light, he fancied. Fionn smiled, rubbed his eyes, and passed his fingers through his hair a couple of times. At first Fionn looked away from the camera. Lev looked in and out of the crosshairs of the lens, while making jokes to try and make Fionn feel comfortable, despite the insecurity of his own fingers.

Fionn's face was stealing. A vortex within the tranquility of the scenery.

"Has anyone ever found a constellation in those freckles?" asked Lev.

A face is a code, a shorthand to decipher. Succinct as it is, it is a sizeable piece of information about someone. The human ability to perceive a face is particularly refined and is linked to processes of emotional and cognitive evaluation. When a face is directly gazing in our direction, the more attractive it is, the more pronounced is a response of gratification in us, with an underlying involvement of brain areas that process reward and a concomitant rush of the neurotransmitter dopamine. By looking at a face, we involuntarily form inklings about traits and attitudes. One of the immediate features we infer is whether a person is trustworthy or not. Should we yield to them or should we avoid them? Should we dismiss or trust in their intentions? The process is subliminal. Aided by experience and predictions, we produce those rulings unconsciously, within milliseconds. We are so trained that this happens even when we are exposed to a face for a time so short we don't even realize we have seen it.

As he posed on the bench, Fionn felt the urge to take pictures of Lev in return. It was strange for both of them to linger for the first time on each other's glances, to yield to such an open cross-scrutiny.

"How do you manage to be so together?" Fionn asked Lev all of a sudden.

"What gave you that impression?"

In his book about photography entitled *Camera Lucida*, the French intellectual Roland Barthes has something to say about portraits. "The portrait photograph is a closed field of forces," he says, in which "four image-repertoires intersect . . . oppose and distort each other." When we are in front of a lens, we are at the same time the person we think we are, the person we would like others to think we are, the person the photographer thinks we are, as well as the person the photographer exploits for his or her artistic ambition.

When two people meet and fall in love, they often have jarred impressions of each other. It is possible that we incur a quid pro quo of projecting into the other what we don't possess ourselves, in order to compensate for what we are not.

Since the very first day he met him, Fionn meant lightness to Lev. To Fionn, Lev symbolized strength. In one way or another, they both thought they were exemplars of resolve, the closest it got to a vision of what it means to be a decent man.

But in the throes of love, the euphoria and the expectations aired by desire muddle with our powers of discernment. In the brain, this is apparent in deactivations in areas that regulate negative emotions, judgments, as well as the perception of oneself in relation to others.

Lev was convinced he knew how to recognize a beautiful being when he saw one. Fionn's biggest wish was to find one. When, unexpectedly, serious beauty crosses our path, the encounter is as exalting as it may be disorienting. It demands truth of us. It is also true that familiarity makes it easier to appreciate an object. For that, gazing at a safe distance never suffices. Beauty wants to be touched and explored. It wants to be stirred, perturbed, unmasked. It wants to echo.

The moment Fionn turned his head to glimpse the water, Lev

grabbed him and obscured the panorama with the whole of himself. Together they were captured in an intimacy that later turned out to be inescapable.

*   *   *

After a few years spent working for an Israeli newspaper, Lev went to New York to teach creative writing part time in an art school so he could pay his rent while he continued to work on his first novel, which he had been laboring on for the past two years. He was the youngest child after four sisters. His grandparents arrived in Israel from Venice, though for dozens of generations before that the family had lived in Sicily. Fionn came from Bantry in West Cork and grew up talking to horses in a farm at the edge of a peninsula, together with his younger brother, Tom. Born on the same day eight years apart, Lev and Fionn later found out they had also arrived in New York on the same day. Fionn's dream was to work in journalism and broadcasting. He loved radio. A student ever since he left Ireland, as a migrant Fionn was torn between the opportunities of a future abroad and the embrace of a return home. To everyone who asked him how he liked living in New York, he said: "I'm only here for a short spell. I'll soon go back." Lev loathed academia, but he was passionate about teaching. In the classroom, his enthusiasm was arresting and made students feel like protagonists of a revolution. Doing all he could to try not to make it seem obvious, Fionn, who was in a different school and had taken Lev's course as an extra unit, found excuses to stand or sit next to him in the classroom. He respected Lev and was enticed by his vision and by his propriety. Fionn wanted to be introduced to Lev's friends—who were all older, even than Lev—and hear all the things he knew.

After the afternoon by the water, the two became inseparable.

Lightness. The word "lightness," and a feeling of peace attached to it, recurred in Lev's mind at the thought of Fionn. Lev was unrecognizable when he was in love. When his proclivities were requited, he became a pulling force and brimmed with fearlessness. His seemingly poised semblance vanished, uncovering an unrestrained core. He did daring stuff, like singing for Fionn from underneath his window and persuading passersby to tell Fionn how handsome he was. In Fionn, Lev found the brother he never had. "There's no bad bone in this boy," he told his best friends. "Everything in him is so instinctual, so unrehearsed."

Lev took Fionn to the Galilean hills, made him wear a kippah at a Pesach seder, and took him to the *giudecca* in Sicily that his family had supposedly come from. He pointed at the spot on the Lebanese border where he had stood with a rifle in his hands and counted the days to the end of his military service, and walked with him on the squares where he had marched and rallied against the occupation of Palestine. Fionn drove Lev to West Cork, introduced him to his family and to his best friend, Shane, and proudly made him ride his horse, Tristan. His first time on a horse, Lev shivered and was embarrassed to show that he wasn't at all comfortable up there.

"Oh, you're grand. You don't need to have passed an exam for this," said Fionn, finding Lev's clumsiness rather cute. "Pretend you are me up there. Tristan will feel that."

Together they used beauty as an excuse for everything. So they drove miles to reach the most scenic coast and watch a sunset. Fionn would persuade Lev to scratch his shins as they passed through thorny bushes to find the best spot for a dive and not to be afraid of climbing a tiny cliff. They would cycle the narrowest roads to notice every hue of the color green, run to follow a dolphin along the pier. Together they would take adventurous flights with their imaginations, discussing their work or responding with wonder to whatever they stumbled upon. So the sight of an old shipwreck in a corner of a bay might have led them all the way back to the arrival of the Vikings. Vague assonance between words in Irish and Hebrew would have made them revive the old speculation that the Irish people were one of the lost tribes of Israel and that, therefore, they had been long related. While edging along a rock, they might have come up with the perfect plot for a short story. Lev would have hinted at choices for a scene, and Fionn, in return, would have imagined how the story would hypnotize listeners on radio. So they enjoyed marveling. Neither of them had had this with anybody else before. Give or take, they were similar in their unwillingness to yield to focus and exactness. On such journeys, they would soar, holding each other's hand, except Lev would occasionally become afraid of heights and beg to land again and park their fantasy.

On their last summer night in West Cork, they watched the day end on Sheep's Head, the very tip of the peninsula.

"Do we really have to go back over there tomorrow?" asked Fionn, pointing to the other side of the Atlantic.

"Not if we don't want to . . . , but what about your degree?"

"Listen," said Fionn, "one day we'll both move here. We'll make this happen . . . I know I'd rush back home, if you followed me. I miss home."

They could never decide whether Hebrew or Irish owned the sweetest sounds to say "good night": *layla tov* or *oíche mhaith (ee-he wha)*.

Unexpectedly one night, after they'd made love and Fionn was already asleep, Lev broke into a sobbing cry. He didn't know yet where those tears came from.

*　　*　　*

Their first birthday together arrived.

"Is there a name for us?"

"What do you mean?"

"For those whose birthday is on the same day?"

"I don't know," said Fionn. "Why don't we just make one up?"

Fionn gave Lev a rare first edition he had sighted months before in an antique bookshop and also left him a sweet voice memo on the computer. Fionn always chose the perfect gifts, effortlessly. When he was a child, there wasn't a present that wouldn't make him jump from joy, like the little radio his dad had built with pieces from an electronics shop. His parents couldn't always afford the expensive toys he had heard about in school, but he never complained about what he was given. Lev got Fionn a book, music, and a hat, because he couldn't decide what to pick. Fionn's parents, Patrick and Nancy, were a force. For the love of her husband, Nancy, who was American, had to give up her family, who disowned her for having chosen a man they didn't approve of. A stranger in a new land, she became a source of strength for all the many close friends she made in Ireland, and her love for her husband and two sons became her deepest foundation. This engraved in Fionn the notion that in the absence of all other certainties, love, as uncertain as it is, is still the safest bet. Lev had always been a parent for his own parents. He always had to reassure them he was not neglecting them and that he was investing in his professional future, which to them counted more than the private one. When, as a child, Lev brought friends home for his birthday, his parents cleaned the house thoroughly, bought them cake, and cooked them dinner but weren't interested in remembering their names. They even showed some jealousy toward their son's friends and occasionally asked Lev to skip his playdates to stay with them, because there was a relative to visit or a family function to attend. Over the years, Lev had to learn not to yield to his parents' requests, something he struggled with even as an adult. Lev was old when he was young. As he aged, he aspired to be younger. This aspiration made him somewhat clumsy in his manner.

Lev was giving and, yes, did he love, but in a way one could say was always somewhat selfish, as if the act of loving counted more than the person to whom he directed his sentiment, as if gestures followed a formal protocol rather than a free inclination. Questions lurked behind his expansiveness: Is this feeling right? Am I behaving right? What would others think? This was probably why even with Fionn Lev always sought confirmation for feelings—his own and Fionn's. He wanted to articulate meaning from their actions, frame them inside a structure. He looked for rules. And he felt embarrassed for that.

Fionn'd say: "Relax! Can't we just live the stuff that happens to us?"

So Lev would give himself over. He would break out of inhibitions. But then something inside him, a diffidence of a sort, hastened to suggest to him he withdraw before he could fully let himself go, as if he wasn't allowed, or as if anticipating failure or pain, he needed to protect himself from it. This tendency would manifest in different aspects of his life and made Lev occasionally neglect to enjoy the moment and simple pleasures. It could happen that even during their most spontaneous initiatives, a "but" would occasionally surface, an adversative reaction that inevitably killed the initial enthusiasm. An abortion of impulse, leaving those around him surprised by the sudden turn and believing Lev acted out of obligation. Slowly, Fionn began to notice this, which terrified Lev.

In a relationship, much of what we may be scared to be judged for is random. Body size, a poor job, taste in music, bad breath, snoring, hypochondria, chronic untidiness, ignorance of modern art. But more frightening revelations are typically linked to inner wounds. These are painful chapters in our emotional biographies, regrettable actions or mishaps in relationships that if disclosed, we dread, would keep partners away. Difficult parents, a complicated childhood. Maybe the fact that we have never been able to keep a partner (or that we have had too many), that we have been divorced several times, that we have hurt or let people hurt us, or in some very unfortunate cases that we have been victims of abuse or violence. All this can even make us feel shame. Shame is a toxic emotion. It lurks silently inside. The less it is shared or exposed, the stronger it grows. It's a poison with a long half-life.

Around the time Lev met Fionn, he was not entirely at ease with himself.

First, there was a question of accomplishment. Torn between a career in academia and one as a writer, he muddled his ambitions. Younger and less experienced, Fionn had clearer ideas about his future. Tellingly, Fionn also didn't fret if his prospects for the future occasionally clouded. "I'll figure it out," he'd say.

Something else made Lev insecure. The men Lev was typically attracted to had one thing in common with Fionn: they all seemed unconcerned, self-assured, at ease with themselves. Light. Because Lev was self-conscious about being heavy in comparison. Like Fionn, they were drawn at first to Lev's worldliness, to his captivating intelligence, to his wisdom, and to his maturity. But then Lev punctually showed his edges. And he did something else. Unfailingly, he made lovers feel they were the exclusive source of his joy, the reparation for his faults, or that they needed to give him their approval to win his sentiments. Unfailingly, they didn't stick. On his way out, one of them once told him: "First go figure out whether you want to lead or be led!"

With Fionn, he made the same mistake. He believed the boy to be the antidote to the heaviness he perceived in himself. The only difference was that Lev sensed Fionn was precious and different from all the others, and that Fionn seemed not to want to flee anywhere. Lev was in awe of Fionn, and of the strong bond that was growing between them. Instead of elating him, this thought secretly terrified him. Lev wanted to get it right this time. Occupied as he was with his own conjectures, he neglected what Fionn really had to give. Unassuming, disarming, Fionn had shown up like a gift. The offering, in all its expansiveness, disoriented Lev.

* * *

Fionn was visiting from New York. Lev had lived alone in a cottage for about four months. After they had been together for almost one year, Lev quit his job and took a writing bursary in the Irish countryside. Fionn was beyond happy and excited for Lev, and told the news to anybody he knew, his parents first. He was still to finish up his degree in New York, so for him Lev's move also meant one of them had made it *home*. Lev had decided on the spur of the moment without knowing really why he was going, but more for Fionn than for himself.

"Can't you come in a couple of weeks?" Lev had asked. "That would be better. I'll have more time."

"No, this is when I can take time off my classes."

So, Fionn had got on the plane.

"I care for you so . . . I could beat someone up if they hurt you, is all," whispered Fionn, Lev in his arms, as they fell asleep the night before a mountain excursion with Tristan. "I really could . . . Please stay here, and I'll join you in a year or so after I finish that fucking degree. You and Tristan can become best friends in the meantime."

"You know, there's no future tense in Sicilian . . . I'd better make no promises."

In the morning Lev woke up from a bad dream in which he saw Fionn approach him with a smile, but he couldn't smile back. Lev forced himself to beam, but he just couldn't bring himself to do it. His lips wouldn't arch, and Fionn vanished in tears.

Lev had not had much of a social life in his new surroundings.

His only outings were a few beers with his neighbors, walks to the sea or the supermarket, and visits to Fionn's parents, who grew fond of him. He struggled to get on with his work and many times he was on the verge of giving it all up. Fionn was there, but Lev just couldn't feel his presence, so engrossed he was with his own stuff.

"We don't need a map, man," said Fionn, with a sneer.

"I want to bring one!" snapped Lev.

From the start of their ascent, Lev complained about everything. It was too cold and windy. The path was muddy. The trip would take too long. Would Tristan mind the two of them on his back? What was there to see at the top anyway? And, why didn't they stay home instead? They could have cooked something, and he could have continued to work. When they stopped to take breaks, instead of embracing Fionn's company, Lev sat down with a book, kept checking his watch and his phone, fought with the folds of the map. There wasn't a cliff or a view that Fionn could entice him with, not an excuse that could make them marvel as they normally would. On the contrary, Lev found ways to establish a distance, erect a hierarchy, to test and challenge Fionn, something he had not even done when they were together in a classroom. What did Fionn really know about the history of the place? How did he see his future? Was a career in radio possible? And when was he going to finish his degree? Gratuitously, he made Fionn feel inadequate. Fionn began to be impatient but didn't yield to a tiff.

Fionn had let Lev sit in front. Close to the top, there was one particularly steep passage ahead of them.

"No, we can't do this. Let's take another way."

"Of course we can. I've done it many times. Come on, it's the right path."

"It's too dangerous. We shouldn't do this."

"Go on. Tristan is strong."

"No."

"Trust me. I'm behind you. What could happen?"

Lev dismounted.

"We actually need to start to go back. The people from the bursary are expecting us for dinner. We can't let them down."

"You didn't tell me that."

"I'm telling you now."

"With due respect, who fucking cares about this dinner?"

"I do."

"Listen, I've come all the way here for you. What's your fucking problem, Lev?"

"I have moved here for you."

"Then bloody be here with me for real . . . I've missed you, but it seems whatever I say or do is wrong."

"What do you want from me, Fionn?"

"Nothing. I just don't understand why you are not happy, Lev."

Lev tore the map and threw it on the ground.

"What's important for you, Lev? Will you ever know?"

Lev found himself in a position that scared him to death: He couldn't escape himself. He feared his life had taken a course that he didn't like and was impossible to divert.

A mordant grip on our true inclinations can become a comfortable habit. But eventually the grip will tighten and we will feel trapped. Lev felt like he had literally climbed and climbed a slope and reached a summit that had no foundations. He hadn't found himself. Lev's anxiety, his dejection and his irritability, originated in an obstinate kind of self-perceived dishonesty. Lev had always felt Fionn closest when Fionn felt protected by him, and when he gave Fionn the chance to admire him, to look up to him. Disoriented, empty, and unable to enthrall Fionn with any security, let alone firmness of feeling, which he believed Fionn needed, Lev felt helpless. Yet he didn't know what to do.

Naturally, Fionn sensed Lev's lack of strength, but he didn't expect much, perhaps only that Lev could trust him, and he couldn't understand why his love for Lev just didn't get across. A vexing confusion of expectations and a perverse reciprocity of frustration. Before then, they had relied on undeniable affinities, but obviously something had to give, something deeper had to be resolved.

During their course, at some point relationships inevitably face a mirror. When that happens, we can either look face on or run away. Each choice has its consequences. When we find ourselves unacceptable, when the threat of rejection, concrete or imagined, looms, there is a chance that we will strike out. Lev couldn't believe what he was about to do or why. He had to swallow before he opened his mouth, but he raised his head toward Fionn and said, "We're done . . . I think we must go in different directions," and started to walk back down.

"That's damn right," said Fionn impulsively. And then, himself doubtful and exasperated for the first time, he added: "Maybe certain things will never change."

When people part ways, whatever remains unsaid or undone feels like dust in their veins. Thoughts shake on nerves, syllables balance on the tongue. Kisses perch on lips. Teeth grind. Fists strangle air.

Lev walked without a direction and stopped at a pub or two. Twice his bike struck a parked car and he fell in front of passersby.

"Hurting someone is an act of reluctant intimacy," wrote Hanif Kureishi. Fionn should have been a comfort to cling to, a slate on which to compose the truth about himself, but Lev couldn't see it that way. Lev's happiness had chains and the imprisonment was self-imposed. His was the behavior of someone who expects, almost desires, to be crushed. Fionn didn't know how to help him. Even if his heart was open and capacious, it wasn't bottomless. It's hard to give when those we give to don't know how to receive.

Referring to the construction of an original aerostat that was supposed to fly for the first time over the channel in 1785 and was made of two balloons, a fire balloon for better control and a hydrogen one for greater lift, the author Julian Barnes wrote:

"You put together two people who have not been put together before; and sometimes the world is changed, sometimes not. They may crash and burn, or burn and crash. But sometimes, something new is made, and then the world is changed."

The pairing of Fionn and Lev was an audacity. Their worlds changed, but they didn't immediately understand how and for what reason. Light and heavy tried to mix. In the process, they lost altitude, swirled, and got sucked into air pockets until they were able to glide again and recuperate thrust. At once allies and competitors, they needed time to understand their own purpose in order to make space for the other.

Differences. Gradients. Imbalance. Incommensurability. Lev wasn't ready and, in truth, maybe neither was Fionn.

When Lev returned to the cottage, Fionn was gone. Naked in the shower, Lev shouted, then shriveled, sat down under the running water, and fell asleep.

Before flying back to New York, Fionn went to see his family and rode Tristan one more time at night.

* * *

Three years went by. One to hide, the second to begin to breathe, and the third to rise again. Apart, but carrying the constant memory of each other, the two lumbered, more or less defenseless, more or less pliable to life's offers and requests. As Patti Smith wrote of her relationship with Robert Mapplethorpe, they went their separate ways, "but within walking distance of one another."

Fionn was hurt, but he had done nothing wrong, he thought. Every now and then, he would send Lev a postcard. One came from Israel. There, Alana, an archaeologist from America who looked for Roman traces in the occupied territories, loved him, and he let himself be loved for a while, to deal with the doubt of what might have been if he hadn't let Lev go. Lev never replied, and Fionn stopped writing.

Lev began to accept the fact that for his own sake he needed to stop letting his past condition his future. So during those three years Lev worked toward change.

Change is hard in all aspects of our lives and especially so when it concerns the way we love. When we aim for change, we must recognize the need for it, then come to desire it with such obstinacy we become prepared to make costly sacrifices. The priciest is to find the courage to face ourselves for what we are and not look away, however broken or repelling that view may seem to us. Lev had a lot to sort out for himself. After his novel was finally published, he resolved that life as a writer didn't suit him. The solitude it required, the self-obsession it generated, and the absorption into an imaginary world was too alienating for him. He didn't know how to prevent that regimen from getting in the way of his personal life. Most writers don't choose to write. Writing chooses them. Despite his talent for it, Lev's passion for writing wasn't inevitable. Writing made him available to nobody but himself, and he couldn't afford that.

The prohibitions we build against our intimate fulfillment come in different forms and originate in all kinds of unresolved issues. Without realizing it, we fall into self-created traps from which it's difficult to escape. We may unconsciously make choices, one after the other, that create a dissonance between who we are and how we live, what we want and what we can offer, what we desire and what we ask for, so that we protect ourselves from showing and getting in touch with our raw selves. In one way or another, we stubbornly perpetuate situations that keep us safe.

Entangled by his own chains, when Lev first met Fionn he was oblivious to his contortions. He couldn't hear himself scream. But then something screeched. Now tormented by the loss of Fionn and the prospect of his never returning or him never finding anyone else, Lev shouted, "Enough!" and began to make different choices. He worked hard to get a teaching job again, eventually finding one in Dublin, because it was in teaching that he was in his element. He ceased being too focused on himself, even less on what others might have thought of his choices. He resolved not to spend a day alone. He invested time in new friendships. He took the walks he had always neglected to take. Whenever he was worried about something, he called friends to ask them how *they* were doing. He took riding lessons.

On a path toward change, we embark on new roads, take turns, sometimes traverse dark tunnels and emerge in uncharted territories. These are all tracks that contribute to shaping who we become. One change paves the way for another, so a small step can overturn a whole catalog of undesired habits. We learn by doing. We all deserve to find what it is that, more than anything else, matters to us and makes us vibrate with life. We need to be authentic. Already at the end of the nineteenth century, William James had recognized that change could be initiated by a shift of mental habits. He wrote: "Seek out that particular mental attribute which makes you feel most deeply and vitally alive, along with which comes the inner voice which says, 'This is the real me,' and when you have found that attitude, follow it." In neuroscience, this translates into small adjustments in the way neurons fire that, repetition after repetition, have the power to make us come out of a habitual rut of behavior and settle on a better track, shifting, for instance, automated responses of fear to attitudes of positivity, or inaction into purpose. Habits respond to cues that trigger them. Over time, habits ossify and become so encrusted in behavioral riffs because, in one way or another, they reward us. In order to break habits, we need to recognize those cues and avoid them or force ourselves to respond to them differently, to experiment with new rewards.

Lev went from one boyfriend to another. They were all only pastimes, bridges that made Lev reach his next point in a better vision of himself. James, a kind banker, was the one who lasted the longest, but, feeling the specter of Fionn, did Lev the courtesy of bowing out, to let Lev benefit from the sting of a story left undone and perhaps mend it.

Lev came to accept his limits. He understood that a suitable person for him would only be someone whose reservoir of acceptance and benevolence, will and generous circumstances of life, exceeded his. Someone who could compensate for his instability. Fionn was that and, deep inside, Lev had known it all along. But when he was around Fionn, he just wasn't ready to admit it. He wasn't ready to admit his own vulnerability and made himself unbearable as a result. This is a crucial point. We don't need to be perfect to be included in someone else's life. What counts is to see that we are not perfect and to be willing to accept it. That's the trick. When that's in place, when we are ready to be ourselves, only then can we become generous enough, to others and ourselves, that no rejection can hurt us. Love does not arise from protection. It grows best in the rawness of vulnerability. Love is to submerge fear with courage. It's to efface ourselves in a way that doesn't erase us but makes us overcome unnecessary insecurities and overlook superfluous needs to be able to attend to those of the beloved. Lev wasn't really ready to receive. He needed to accept that he deserved receiving and that good could come out of it. Accept that he could be loved. He needed to put his lover, not the danger associated to yielding to him, at the center of his attention. That's where he could learn how to stop his inconvenient habits. And that's when his hunger could turn into generosity. All he needed was to reverse the attention: listen to what others had to offer and needed, rather than being preoccupied with how he loved them or if they loved him back. After three years, Lev didn't become a different person but someone who at least knew himself better. That awareness gave him calm and made him look more at peace with himself.

Fionn was the best thing that had ever happened to him, the man who inspired truth in himself, and Lev now could say it without fear, without shame. And with gratitude, because Fionn pointed to what really mattered. Thanks to Fionn, Lev could look himself in the mirror. When two people realize they can't live without each other, it's because they recognize and need the power of that mirror. When that happens, they become inseparable. Learning to trust oneself and others can be a tortuous path, but it leads to this richness.

It's easy to think that just because intimacy failed once or didn't catch on as imagined, it will never happen or that two people will never be in a position to share it again. No. Intimacy quenches and fades. It can hide, but it can reappear vigorously in one moment. Lev and Fionn never stopped being intimate.

Now Lev's greatest wish was to learn how to teach and be taught. And he wanted this with Fionn. There was no ranking anymore, no comparison, no judgment. There was no question of leading or being led. Together lightness and heaviness made depth. They were *inter pares*. They were equal.

* * *

Lev turns up the volume on the radio, although he knows instantly. It hits him like a wave.

Fionn's home.

Like a face, a voice summons our abilities to distinguish. It marks distinct traces in our memory and its recognition has a dedicated neural repository in the brain.

In the 1920s the biological geography of a memory was given the name "engram"—it sounds like something between software and a groove in vinyl. Since then, through decades of surveying, which involved erasure and manipulation, it's been possible to locate engrams and thus map memories across the brain. Even within small clusters of neurons. Most recently it's been possible to selectively recall memories. In a study with rodents, researchers were able to specifically revive stored memories of enjoyable episodes and use them to steer behavior. These results pave the way for the possibility of treatments for conditions like depression, in which those who suffer from it benefit from remembering pleasant rather than negative experiences.

We could all benefit from capitalizing on positive memories and using them as springboards for our mood and for actions. Lev has left darkness behind and only wants to cling to those moments when nothing, nobody, could have denied his tie to Fionn. And those moments were many.

Listening to Fionn's voice, he sees Fionn's incredulous eyes when he said yes to the concert invitation. He hears Fionn's laughter in the voice memos he sent him for his birthday. Lev's memory travels to the night when they got drunk at the bar and flirted with waiters. And to when Fionn got jealous after he spoke to another guy at the supermarket. To that morning when, after all the students including Fionn had been stranded on Lev's floor because of a storm at the end of a gathering, Lev woke up and found Fionn wrapped up in clothes he had taken from Lev's closet during the night. To when Fionn used "us" instead of "me" for the first time. To the time Fionn was the only one who asked Lev if everything was all right after Lev had to deal with a hysterical head of department. To when, sitting on a bench, they planned their first dinner party together. To a room full of guests in which Fionn regularly looked at Lev from one corner to simply signal that he was there, and then laughed forbearingly when he saw that Lev, as usual, had cooked too much food. To the waltz. To when, on a boat, leg to leg and an oar each, they rowed until it got dark. And to when Fionn made sure Lev knew that he had understood it was excitement, not nervousness, that often made Lev look like he was fidgeting.

These rich memories, and many more, elevate Lev to a height from which nothing, no doubt, could make him descend. Such are the acrobatics of love. He knows he never wanted to hurt Fionn and that he will never hurt him again.

Lev's wiser, defiant, fortified. His heart's for Fionn.

Erratically, during the previous three years, whenever Lev needed to talk to Fionn, he wrote letters he never sent. They contained no question, no explanation, no request. Only a chronicle of the everyday. They were the journal of Lev's growth. Fionn would probably read them, Lev hopes. In any case, there is no time to ponder or hesitate. Lev needed to send them to Fionn at once.

*   *   *

The day of the referendum, Fionn returned to Bantry, where he was registered to vote. First he went to the polls, then rode Tristan with his brother, Tom, to Sheep's Head and back. Later he had a pint with his buddy Shane. After dinner, he sat down on the grass with his ma and da, looking at the colors drift on the bay. There was peace all around, only the cuddling sound of waves splitting on the rocks beneath them.

"You're home now," his ma said.

"Yeah, it feels great."

"Tristan is happy to see you, apparently," said his da. "I remember when he was still a foal and got very sick. You didn't separate from him a second until he got better."

"We're confident about the vote tomorrow, you know. . . . We really are," his ma said.

"I believe in it too," said Fionn. "What's happening is amazing."

Then there was a moment of silence.

"I hope you don't mind us saying this, but there might be something else going through your mind these days, is it?" said his da.

"What so?"

"We know Lev's still here."

"He doesn't talk to me," said Fionn, and stood up.

Then he said: "I'm afraid."

"You, afraid? That's new. Of what?" his ma asked.

"That it might have all faded."

"Did it fade for you?" she asked.

"No, but he wouldn't hear of me."

"Fionn, come here." She took his hand.

"Listen, I know he hurt you, but don't you think he still cares a lot about you?"

Fionn closed his eyes.

"Fionn, do you remember the time when your little brother used to be bullied and stopped playing with you because he secretly wanted to be like you, but . . . he couldn't, and you refused to play with anybody else until you made him understand there was so much to him he didn't know?" she asked.

"So . . ."

She continued: "You have never let anybody discourage you from loving them."

Fionn gripped his ma's hand and reached for his da's to join theirs.

"You see, we're made to care, no matter what it takes," she said.

"I saw you and Lev together in the best of times. 'He's such a great man,' you used to say, 'so fair, so decent.' Certain things don't fade that easily," said his da.

"What am I supposed to do now? I tried. He didn't give me a chance."

At this point, his ma said: "This morning a parcel arrived for you, from Lev."

Tomorrow was a chance, and hope made night last a second.

One after the other, Fionn devoured the letters. In the morning, he took the earliest bus and then a train to Dublin. By the time he got there, the tallies started to come in from around the country. He ran his program before he went to watch the announcement of the results at Dublin Castle. There, the sky unbreakable and the sun giving on a May Saturday afternoon, everyone hugged, shivered, beamed, and cried. The nation had voted in favor of making same-sex marriage legal. There were the young and the old, the single, children, teenagers, parents and grandparents, gay and straight, lovers, friends and enemies. Songs were sung. Cheering firemen did the rounds up and down in their trucks on Dame Street. That "yes" was an elevation. A gift beyond tolerance. A reassurance many couldn't have given themselves alone but only the generosity of a large portion of the nation could have granted: that their sentiments are noble, that they are protected, and that they are all equal. All equal.

Lev and Fionn sought out and found each other again among the crowd. The moment they were in front of each other, Lev grabbed

Fionn by the hair and pulled him against himself, like you would snatch a child from the way of imminent danger. Twice he uttered some version of "I'm yours," and twice he said "Sorry" into Fionn's ear and neck.

It was repentance, devotion, forgiveness, protection, generosity, and recompense.

It was equality.

Yes

It feels like he is here, watching. I could turn around and see him in bed, leaning on one elbow, asking me why the hell I can never sleep, if I know Jack Spicer's poem about one-night stands, and whether I could rub his back. He knows he gave me a hard assignment. I wonder, did I take it too seriously, or did I fail to fulfill it?

You must be home by now, Anthony, unaware of what you have unearthed by simply being who you are. Not one week since I heard you whistle, and it's simply a wonder our paths crossed. Through you, Maurice has vigorously reemerged from wherever he was to remind me of what he and I decided our lives were going to be.

Don't be upset that I ran away earlier. I was overwhelmed by your resemblance to him. It is time I told you about him, as it seems to me the two of you entertain some secret relationship.

Tonight your eyes looked almost like his. You are fairer. His hands were larger. Your neck is slightly wider. But air whispers the same way when you are around, as though something were about to be revealed. And the smile . . . the unwavering curl on the lips and the eyes thinning when you are about to laugh—these seem to have traveled straight from his face and landed comfortably on yours. I wrap my legs around you in a body signature that was ours, Maurice's and mine. I have to remember to avoid your knees, because you are not him, after all. He was exquisitely sensitive on his calves, and to how I used to trace them with my shin, in a chaining caress. If he really were here now, he would touch the back of my neck. Bury his nose in my hair. Kiss my forehead.

"It's all right, Margo. It's all right," he would say, "Together we figured out how to live—don't you remember that?"

If neither of us had brought anyone home, we would spend the night together. He would emerge from his room in the basement and slip into our big bed upstairs, without saying a word, sometimes humming a tune. Then it would start. A soft, shared murmuring on the day just gone. I would ask him things like "Was it hard to better the world today?" And he would say, "It was enough to step outside and show up there . . ."

It was our lullaby. In the middle of it, he would pull me aside and ask, looking me straight in the eyes, what was wrong with all the lovers in town who didn't know I existed. But then, with a tinge of fear in his eyes, he would ask: "You are not letting anyone take my place, though, are you?" as he nosed my cheeks with his eyes closed.

"No . . . No one takes anyone's place here," I would reassure him. "This is fairies' heaven. There's plenty for everyone."

This is what bedtime used to be at our place, especially after our promise.

* * *

Being touched is never only a passive experience. It involves being engulfed in someone else's movements and intentions. Two bodies that interlace feel united and at the same time realize they are separate. Being touched has the subtle power to awaken parts of the body that might otherwise seem dead or forgotten. It inscribes meaning on them. Touch is delicate yet powerful at communicating emotions. It does so with tremendous nuance. Touch has form, speed, duration, intensity, and sway. It can be firm and gentle. It advances and halts. It spreads and narrows. Like a kind of handwriting, touch spills the ink of emotions with sophistication and, despite being entirely personal, does so unambiguously. This is exceptionally true for two people who are well acquainted with each other's body habits, but it can also occur between strangers. One elegant study showed that randomly assigned pairs of people accurately communicated to each other emotions such as joy, anger, sadness, fear, love, disgust, sympathy, and gratitude through touch.

By patting, stroking, rubbing, pinching, squeezing, or nuzzling, the initiators of the touch dialogue conveyed an emotion of their choice by making physical contact with the recipient partners, who, blindfolded, guessed what was being communicated to them. A code of tactile choice, duration, and intensity emerged from the study. For instance, sympathy was mostly conveyed via patting and rubbing. Anger took the form of shakes, pushes, and squeezes of strong intensity and short duration. Hugs, strokes, and nuzzles of sadness were of lighter intensity, but their duration was on average longer.

From lines in the skin to vaults in the brain, the neurological architecture underlying touch is being unlocked. Pleasant soft and gentle strokes on hairy skin stimulate specific nerve endings that project all the way to the insular cortex, a corner in the brain associated with the elaboration of positive emotions.

Roland Barthes said that "language is a skin" and can shake "with desire." "I rub my language," he writes, "against the other." It's like having words tripping off fingers "or fingers at the tip of my words." Lovers tilt at either side of the equation, practicing their own dialects, inventing new ones. The body becomes a writing surface; touch has its own dictionary.

*　　*　　*

Margo first saw Maurice's deep black eyes at a party. His white-and-blue striped shirt open, a thin red scarf around his neck.

"What's your passion?" Maurice asked her.

With this opening line, Maurice tried to glimpse everyone's true desires. He wanted to know what kept them alive. He counted the seconds until he got an answer. If he had to wait long or if the answer was evasive, he would move on to the next target, in search of better enticement.

"Sex, really. Nothing else" was Margo's reply.

She exaggerated. In fact, her sexual world developed only after meeting Maurice, but she knew he would like such an answer.

"Hey, what's yours, then?" she challenged him.

"To help others find and express theirs," Maurice replied.

Only later did Maurice ask her name. He said he had continued to converse with her because he found her hopelessly doomed to a complicated life. That was the kind of misfortune in a person he was willing to share.

"Because life is sweetly complicated when you decide that love is its premise," he said, bringing a glass to his mouth.

They walked to the sea and talked until the morning seagulls returned from the horizon. He then carried Margo all the way to her room. He had powerful, supple hands with large, rounded knuckles. The fingers long, in agreement with his ability to discriminate detail.

As he said good-bye, he shouted: "I think I have a new passion, Margo!"

He ended on the *o* as if he were blowing a kiss.

They moved in together three weeks later, on a bright October afternoon. The air was crisp. On the avenue, the trees bowed with delight as they drove with the loaded truck toward their new little home.

"I plan on bringing many guys home," Maurice warned Margo.

"Fine, I'll just have to approve of each of them!"

*   *   *

People mistook them for brother and sister, perhaps because they perceived in them the same desire to leave their families behind. Maurice was born in the back of his family car somewhere along Highway 1, as his parents drove back from a trip to the woods. As a child, he thought he had invented the word "cuddle" and went around showing everyone what it meant, something he still did when Margo met him.

When he arrived in town, he had given up a prestigious fellowship in law to devote himself to studying literature. His strict immigrant Greek parents refused to speak to him after he left home. The only one he really missed was his mother, Maria. Margo read it in his eyes each time they cooked eggplant. At some point Maria broke the silence and started to visit him occasionally, without telling her husband. She used to bring with her a few bottles of homemade tomato sauce and a book, so Maurice would never forget the country and language of his parents. Inside the book, randomly sandwiched between the pages, would be a small envelope with some of her savings. Both Margo and Maurice had jobs, but money never abounded. Maurice worked as a cashier in a supermarket five blocks away from where they lived. He was there mostly during the afternoons, and his earnings were just enough to pay the rent and keep a bottle of gin and cigarettes at home—which they called their bread and butter. Margo worked in a flower shop. On her free mornings, when she didn't have to rise that early, he would crawl out of bed and go to their big kitchen, where she would brew him coffee while he did push-ups.

Then it was time for a smoke. Margo didn't smoke before she met Maurice, and he put it like this: "You are teaching me how to live. I'm teaching you how to die!"

They had a large chalkboard in the kitchen and they never used it to draw up their shopping list. Maurice would sketch a little something on it or scribble a clue referring to a poem or story he had read. He loved to keep her guessing. He left a question mark and some space for her to write the title. Otherwise, one of them would start a big picture in colored chalk. The other would add, erase, alter . . . back and forth. A rooster meant a boy was in the house. A sailboat that they were going out. A half-moon meant a quiet night for just the two of them. A cloud signaled those rare moments when one of them needed to be left alone.

A large, old wooden desk standing in a corner by the front window was the true alcove of their intimacy.

"There you go, beauty," he said, as he spread the second layer of white paint on it. "Now we can live in here!"

Their worlds fused at the old desk, where their papers mixed. Maurice started the day with her words. As Margo rinsed flowers in the shop, she thought, in trepidation, of him, scruffy hair and semi-closed eyes, reading her words with curiosity. Maurice would then write his own lines and leave them there for Margo to find. Maurice was very disciplined. He wrote every day, especially poems, and he wouldn't leave the desk unless he was satisfied with his composition.

* * *

Margo listened to music when she went to bed. This habit started with Maurice, but it went back to her childhood. She lived in a big house. One of her uncles lived there too. When she was little, she had a room to herself. Her uncle liked to watch TV until late. If she could hear the television, she knew he was occupied and it was safe to sleep.

She taught herself to wake whenever the sound from downstairs stopped. Silence at night meant trouble was making its slow way up to her room.

One night, in the house she shared with Maurice, Margo couldn't sleep. Although she was absolutely still, Maurice was aware of her being up. He pretended to sleep deeply. Margo was beside him, watching him, when he suddenly opened one eye.

"You can't fool me. You haven't slept at all. What's wrong?" Then he tried: "What do you hear?"

"Silence, Maurice. I hear silence."

"Silence . . . And that's what scares you? What if I snored?"

He went downstairs and turned on the radio in the kitchen. Then he came back and snuggled up against her while she drifted off to sleep.

Hard experiences in life weave intricate knots in our memory. They insinuate themselves into our bodies and our minds, leaving scars that perturb our well-being and delineate subsequent patterns and habits of intimacy. But the scars of trauma are not always deep and indelible, because we all have different resources to heal them. Such resources are sculpted by several factors, from genes to brain networks and life conditions, which together may compensate for the impact of trauma and favor resistance to it. The experience of trauma has the power to affect the size of a couple of brain regions. One of them, the anterior cingulate cortex, is at the front of the brain. It's involved in the processing of physical and emotional pain, decision making, as well as other human social interactions, such as empathic responses. The other one, nested in the limbic system, is the hippocampus. The hippocampus is where memories are stored and retrieved but also where they can be extinguished. A study comparing the brains of trauma victims who continue to be haunted by memories with brains of trauma victims who do not has revealed that the lack of resilience in the former may depend on deficits in the integrity of the cingulum, a group of brain-matter fibers that connect the cingulate cortex to the hippocampus. An additional role of the anterior cingulate cortex is to aid the hippocampus in the extinction of fear responses, so deterioration of their connection may be what lies behind the persistence of disturbing memories.

One crucial factor in overcoming trauma is to receive meaningful help. The caress of benevolent care as well as social and emotional support are paramount, especially if they come from family or friends. We can be extremely fragile in the wake of trauma but also supple at overcoming it. Maurice was a pillar of strength for Margo. Maurice knew that and protected her.

Ever since that night, falling asleep to the sound of something was for Margo like being held in Maurice's arms.

*　　*　　*

Maurice hummed well but sang terribly, so he preferred Margo to sing for him instead, mostly old songs from the thirties that would get them dancing. Margo sang to him especially while he took a shower or when he groomed in front of the mirror before going out. Maurice was lusty. He landed on their big bed once acknowledging that his diaries read like a whorehouse ledger. Maurice insisted Margo share even this part of his life, so they often wound up at bars and bathhouses together. Nick, an old friend of Maurice's from home, was his partner in crime during these amusements. Maurice had become so popular in one of the bars that the owners would ring the house if he didn't show up. A few guys were selected to follow him home and deepen the acquaintance. Margo would get to meet them at breakfast.

On the inside of their door, they had a brass plate that read: NMBS, for "no more bullshit." A four-letter slogan to reassure each other they had a plan. Partly an invitation to break with the past, it worked as a daily reminder to do all they could to be happy. It was up there to remind them they couldn't afford wasting a single day on something that wasn't for them or that might threaten the truth of their vocations. They made it a point to keep vivid in their minds what their life goals were and what they entailed. They asked each other questions aimed at making sure whatever they were in was exactly what, as Maurice asked at parties, kept them alive. Together they worked out the nuances of NMBS day after day. The NMBS regimen also had to with their hearts. It was mostly about defending love. It helped them turn away from those they liked but who didn't return their feelings, and from those who for one reason or another were resistant to love.

"No civilians in our queendom!" they said to each other. "We have no time for that!"

Maurice was convinced that love takes the form that works best to hold people, if people allow it.

One particular reason why a lot of people were fascinated with Maurice, and sheltered under the arm of his company, was his ability to help them understand what they wanted. He showed a captivating complicity with every detail about them. Although he was often cheeky, Maurice was not arrogant. He understood that individuals are irreplaceable and so he just encouraged them to be honest with themselves, without letting them feel ordinary when they couldn't. That's why people kept coming to him. Over time Maurice and Margo became good at surrounding themselves with those who thought like them, and they learned how to trust their gut feelings when choosing them. Those they loved were cherished like the last drop of gin in the house.

There were no half intentions: if they said yes, it was YES, with all of themselves.

*   *   *

On a hot midsummer Saturday morning, knocked out by the heat, they remained silent in bed longer than usual. Neither of them had to work.

Then Maurice got up and shouted: "Sweetie, we are having a party tonight . . . Get ready, it's going to be fabulous!"

Maurice asked Margo to take care of the food, while he ran around to invite everybody they knew. He instructed Nick to buy booze and prepare sangria. He came back carrying baskets full of flowers, and they started to decorate the porch with red candles. Margo kept dropping things and not paying the right kind of attention to his enthusiasm.

"I know what it is— It's that boy with a 'woman's face and skin like ivory,' right? I read the lines you have been writing for him. What's his name?" enquired Maurice.

"Steven, and he's got warm, talking eyes colored like river water . . . I think he is one of us," she said, sighing.

Margo had audaciously sent poems and flowers to a boy called Steven, who visited the flower shop regularly, but she had failed to find the courage to sign them or leave her number. The party seemed the right excuse to show him her world, but she hesitated to invite him, because she assumed he would say no. Even if in their realm of affections the larger order counted more, being drawn to a particular individual and depending on their reciprocity did matter, and Maurice had only one way of dealing with it:

"Babe, you'll break many hearts if you insist on protecting only yours and refuse to believe there might be a fragile one on the other side of your efforts . . . NMBS! Please be happy, Margo. Please invite that poor boy to the party—quick! When he shows up, just don't forget to ask him about his passion."

This was one of the numerous simple occasions on which Maurice helped Margo feel alive and brave—as though he was flying her to the other side of the moon.

Margo did invite Steven, and while she hoped he would come, having found the courage to invite him was enough.

The guests arrived all together, a swarm of noisy, relentlessly lively, and refreshing creatures Maurice and Margo had drawn to themselves. Hearts holding hands, lives seeking more life. Nick took care of the music. They all danced. Steven arrived thanking Margo for the poems and saying he wished she had signed the first one she had sent him. Maurice watched them talk from one corner of the room. Before leaving, Steven asked Margo on a date for the following Saturday.

She winked at Nick, who put on Maurice's favorite song.

Margo pulled Maurice by the arm and danced with him to remind him they were the core. Right there, in the middle of their living room, they felt at the center of the world, a world she and Maurice had created.

The party ended almost without them realizing. As usual, Nick was the last to leave. They sat on the porch amid the candle lights.

"Come on, the party can't end just yet. Follow me!" Maurice said.

They ran to the pier, laughing and screaming, until they collapsed on the sand. Margo took Maurice's hands, inviting him to look at the horizon. He leaned against her, with delight clinging to his mouth, and raised his eyes toward hers with a smile that embraced her, the ocean, and the entire sky. His forehead radiated calm, and the chin was tilted just so, to manifest the rebelliousness that had always conquered her and she hoped would never fade.

"Beauty," he said softly, with his face buried in hers, "I want us to trust in the power of siding with life and all its most inconvenient forms of expansiveness. Let's welcome vulnerability and believe it's possible for people to find each other. Everything else is detail. Let's promise each other we won't give up. We won't regret it, I am sure."

It was a perfect summer night with an uncountable number of stars that seemed to have stopped twinkling in order to listen. There was almost no breeze and the gentlest of swells. They could only hear the sea and, in the distance, whales singing to each other.

"Yes, Maurice, I promise you."

There they were, convinced they were beginning to establish the metrics by which they would live. Partners in their belief, they felt the rush of the young and hopeful planting their acorns for the future.

As dawn diffused across the violet sky, Maurice carried her home, like on the night they first met. He put her to bed, with a long kiss on her forehead. Margo held his gaze without saying a word. Maurice covered her eyes with his hands and didn't move until she was deeply asleep. Before lying down next to her, he turned the radio on.

*   *   *

Maurice had begun writing a novel in which he wanted to infuse all his theories on love when Margo noticed unusual skin lesions on his back as he took a shower. Before there had been a wild fever that made him cough and sweat at night and didn't vanish for days. It was the beast.

Around that time, the three demographic categories most affected were gay men, junkies, and Haitians. He made a joke: "What's the hardest thing about telling your mother you have AIDS? Convincing her you're Haitian."

Maurice met Maria at a bar close to their home. First the ritual of the book and tomato sauce.

"Thanks, Mother. Margo and I love your sauce."

"You are still living with that girl? When are you going to marry and move in with the woman of your life?"

"Mum, I am sick."

A burden as heavy as mountains was taken off him as he gave away his terrible news, but instead of rising fresh, he fell into a river of sadness that overflowed from his eyes. Maria knew this would happen to her boy since she heard of the first cases on TV. Unfortunately she ran away.

The new winter arrived and Maurice intensified the pace of his writing. Although they never mentioned it, they knew he had to hurry. They pretended things were normal. The jokes didn't stop, neither did the drawing on the board. They had fewer parties but always a house full of friends. They hummed and sang in the shower. The whispering at night deepened. Unfortunately, Maurice's condition did not cease to worsen.

One morning, he was rushed to the hospital while Margo was at work. Nick found him unconscious and shivering at the desk. Nick called the flower shop and Margo ran home.

It took the most out of her during those fragile moments to be absolutely firm about convincing Maurice he didn't need to stick around to keep her company.

He could barely talk, but he said: "Remember what we decided, sweetie . . . Use your fine emotions to build a world where there's time and room enough for poetry."

Margo nodded and embraced him with the whole of herself.

Then he could leave.

Margo held him until they took him away from her. Back home from the hospital, she poured herself a glass of gin, lit a cigarette, and sat at the desk. Their moon shone in the kitchen. He had taken out a picture of himself as a child playing with a kite on the shore. His papers were there, the front sheet a poem:

Where have you been
all this time?
You show up now,
the ungracious twist
in my last joke.

You let me plot
verse onto the day,
shape hope.
Now you make me eat it all up
until I choke.

Gossip has it things
blossom at the tip of life.
I don't know why,
that tip is every day,

Live or die.

Who's to tell me
it wasn't all a big lie?

Shut up, I don't believe you.

And anyway, it's done.
If it is true, tell no one . . .

A poem broken in the making, like his life, just when it had begun to unfold more or less the way he wanted. Nick was the second
to bow out. Steven and half a dozen of those who were at the party
are dead too. Margo doesn't know how she didn't get it.

Keeping the promise was not always simple. Maurice didn't live long enough to struggle with any temptation to give up. Many times Margo felt alone and fragile, on the verge of yielding.

Since Margo met Maurice, she forced herself to always look for meaning, for closeness, for truthfulness, and for compassion—and refused to expect less of life. NMBS. She understood it was imperative to aim high in the accomplishment of one's intimate life. So every day she needed to cherish and draw on the depth of her sentiments, believe in the sincerity of her needs and intentions, in the capacity of her heart, and master her confidence to share it, despite the challenges, and even if it was uncertain and unconventional. Over the years, Margo noticed, especially among the young, that people became more and more disenchanted with love and more afraid of discovering it and letting themselves go. But she needed to keep the promise. That way she would continue to be surrounded by other like-hearted beings and also be of inspiration to those who didn't have the same experience but maybe needed guidance. The yearning for truth became a gift, not a curse.

What else is there to do? Being generous in the world of intimacy is risky but ultimately rewarding. She went for the rewards. There is no other way of living, she thought.

Margo hears Maurice call her name. She doesn't know how, but he always seems to know the best time to talk.

After his death, Maurice was difficult to replace. When Margo met Anthony, he mended the poetry Maurice embroidered onto life. She saw qualities in him that reminded her of Maurice. Life spilled from every inch of his body. He didn't look away incredulously when someone offered him warmth. He had the undying ability to fall into intimacy because it was there and harmless, and there was no reason to be afraid of it when there was no intention of hurting anybody.

Margo guesses it was a gift straight from fairies' heaven that Anthony sat beside her on a bench and leaned against her with a cheeky face when she smiled at him. She knew she could be herself with Anthony—she saw the oaks. No one had ever made her safe enough for a long sleep after Maurice.

Margo climbs into bed and remembers what it was like to sleep close to Maurice's breath, with his legs wrapped around her when he spooned her. She hears the notes of his murmuring and is reminded of their plan.

If she pays careful attention with her eyes closed, she can hear Maurice humming his opinion on her current world. Maurice would approve of Anthony thoroughly and he wouldn't be jealous. He would rush Margo to invite him over for gin and cigarettes and let him be the one she tells her day to.

"What else are you going to do, babe? He's one of us. If someone so rare like this young man appears in your life, you can only say YES. And when we say yes, we mean it. Right?"

"You're right, Maurice. Damn right," Margo says. "There is nothing else to do but to say YES!"

ACKNOWLEDGMENTS

This book has been long in the making and, as we never stop learning about intimacy, it might as well be not finished and only a taster for future reflections.

During my research, I could count on the excellent service of a few libraries. In Berlin, the library of the Wissenschaftskolleg, and in Dublin, the library of Trinity College. I am particularly grateful to the dlr LexIcon public library in Dún Laoghaire, where I sat many hours, especially toward the completion of the book, and where even the longest days of writing are eased by the congeniality of its staff and the loftiness of its environment.

Allison Lorentzen, my delightful editor at Penguin, provided acute comments on the manuscript and was all along formidable at showing me she understood and cared for my work. The very kind Patrick Nolan kept me going with encouraging enthusiasm. At Penguin, I also wish to thank Diego Nuñez for his professional and punctual assistance.

I am grateful to everyone at Conville and Walsh, in particular Alexander Cochrane, Alexandra McNicoll, and Jake Smith-Bosanquet from the foreign rights team for their unique way of welcoming every new book project that lands on their desks.

I was fortunate to have had repeated chats on intimacy with many people. To mention a few: Noga Arikha, Sue Cahill, Dominique Caillat, Ilaria Cicchetti-Nilsson, Rose-Anne Clermont, Marco Giugliano, Jonas Ihle, David Krippendorf, William Mulligan, Jamie O'Neill, Ida Panicelli, Enza Ragusa, Donna Stonecipher, and Katharina Wiedemann.

Oceans of gratitude to my dearest friend Candace Vogler with whom a priming conversation on intimacy that started in Berlin a decade ago continues to this day, to my enormous luck.

I thank David Halperin, who is like an Auden jukebox, for letting me discover and appreciate the poem "The Lesson," and for insightful dialogues during his residency at the Wissenschaftskolleg zu Berlin.

Special thanks to my delightful generous neighbors in my new home in Ireland, Ken and Hazel Henderson, and Frances Stark, for having spoiled me rotten with encouragement at busiest times and for their great welcoming.

Noga Arikha, Stephanie Brancaforte, William Mulligan, Ida Panicelli, and Donna Stonecipher patiently read drafts of chapters, and I am grateful for their extremely helpful comments and criticism. I also thank Enrico Glerean for his reading of "The Leap."

Without Carrie Kania, my literary agent, this book wouldn't exist. I am as ever grateful for her tireless work, advice, and curiosity, and for her subtle attention to the shifts and turns of my inspiration.

Lots of love to my family.

NOTES AND REFERENCES

Shidduch

6   **matters almost tripled:** M. McPherson, L. Smith-Lovin, and
M. E. Brashears, "Social Isolation in America: Changes in Core
Discussion Networks over Two Decades," *American Sociological Review*
71 (2006): 353–75.

6   **in Europe:** Data from a report from the Office for National Statistics:
http://www.dailymail.co.uk/news/article-2661258/Lonely-Britain-EU
-league-table-shows-dont-know-neighbours-no-one-turn-crisis.html.

6   **exercise, or air pollution:** For a summary of the vast deleterious effects
of loneliness on health, please refer to the following reviews, meta-
analyses, and original articles: J. Holt-Lunstad, T. Smith, M. Baker,
T. Harris, and D. Stephenson, "Loneliness and Social Isolation as Risk
Factors for Mortality: A Meta-Analytic Review," *Perspectives on
Psychological Science* 10, no. 2 (March 2015): 227–37. S. Cacioppo,
J. P. Capitanio, and J. T. Cacioppo, "Toward a Neurology of Loneliness,"
Psychological Bulletin 140 (2014): 1464–504. J. T. Cacioppo and
S. Cacioppo, "Social Relationships and Health: The Toxic Effects of
Perceived Social Isolation," *Social and Personality Psychology Compass* 8
(2014): 58–72. J. Holt-Lunstad, T. Smith, and J. B. Layton, "Social
Relationships and Mortality Risk: A Meta-Analytic Review," *PLOS
Medicine* 7 (2010): e1000316. For a comprehensive book on the
neurobiology of loneliness and human connection, see J. T. Cacioppo

and W. Patrick, *Loneliness: Human Nature and the Need for Social Connection*, Norton, 2008.

6   **encourages sleep disruption:** L. C. Hawkley, K. J. Preacher, and J. T. Cacioppo, "Loneliness Impairs Daytime Functioning but Not Sleep Duration," *Health Psychology* 29 (2010): 124–29.

6   **anxiety, and depression:** J. T. Cacioppo, M. H. Hughes, L. J. Waite, L. C. Hawkley, and R. A. Thisted, "Loneliness as a Specific Risk Factor for Depressive Symptoms: Cross-Sectional and Longitudinal Analyses," *Psychology and Aging* 21 (2006): 140–51. E. K. Adam, L. C. Hawkley, B. M. Kudielka, and J. T. Cacioppo, "Day-to-Day Dynamics of Experience-Cortisol Associations in a Population-Based Sample of Older Adults," *Proceedings of the National Academy of Sciences* 103 (2006): 17058–63. L. C. Hawkley, S. W. Cole, J. P. Capitanio, G. J. Norman, and J. T. Cacioppo, "Effects of Social Isolation on Glucocorticoid Regulation in Social Mammals," *Hormones and Behavior* 62 (2012): 314–23.

6   **the cardiovascular system:** L. C. Hawkley, R. A. Thisted, C. M. Masi, and J. T. Cacioppo, "Loneliness Predicts Increased Blood Pressure: 5-Year Cross-Lagged Analyses in Middle-Aged and Older Adults," *Psychology and Aging* 25 (2010): 132–41.

6   **impairs immune defenses:** S. W. Cole, L. C. Hawkley, J. M. G. Arevalo, and J. T. Cacioppo, "Transcript Origin Analysis Identifies Antigen-Presenting Cells as Primary Targets of Socially Regulated Gene-Expression in Leukocytes," *Proceedings of the National Academy of Sciences* 108 (2011): 3080–85. L. M. Jamerka, C. P. Fagundes, J. Peng, J. M. Bennett, R. Glaser, W. B. Malarkey, and J. K. Kiecolt-Glaser, "Loneliness Promotes Inflammation During Acute Stress," *Psychological Science* 24 (2013): 1089–97.

6   **ultimately to dementia:** R. S. Wilson, K. R. Krueger, S. E. Arnold, J. A. Schneider, J. F. Kelly, L. L. Barnes, et al., "Loneliness and Risk of Alzheimer's Disease," *Archives of General Psychiatry* 64 (2007): 234–40.

7 **short of breath:** A description of pseudo-dyspnea can be found on the website by Dr. Hanna Saadah: http://www.hannasaadah.com/blog/medical/false-shortness-of-breath-pseudo-dyspnea-december-12/.

7 **with the brain:** These are the basic tenets of embodied cognition. For an introduction to intimacy inscribed within an embodied cognition framework, see K. Maclaren, "Intimacy and Embodiment: An Introduction," *Emotion, Space and Society* 13, Special Issue on Intimacy (2014): 55–64.

7 **to social interactions:** B. E. Kok, K. A. Coffey, M. A. Cohn, L. I. Catalino, T. Vacharkullsemsuk, S. B. Algoe, M. Brantley, and B. L. Fredrickson, "How Positive Emotions Build Physical Health: Perceived Positive Social Connections Account for the Upward Spiral between Positive Emotions and Vagal Tone," *Psychological Science* 24 (2013): 1123–32. B. E. Kok and B. L. Fredrickson, "Upward Spirals of the Heart: Autonomic Flexibility, as Indexed by Vagal Tone, Reciprocally and Prospectively Predicts Positive Emotions and Social Connectedness," *Biological Psychology* 85 (2010): 432–36.

7 **our gastrointestinal tract:** L. Dobrek, M. Nowakowski, M. Mazur, R. M. Herman, and P. J. Thor, "Disturbances of the Parasympathetic Branch of the Autonomic Nervous System in Patients with Gastroesophageal Reflux Disease (GERD) Estimated by Short-Term Heart-Rate Variability Recordings," *Journal of Physiology and Pharmacology* 55, suppl. 2 (2004): 77–90. R. Fass and G. Tougas, "Functional Heartburn: The Stimulus, the Pain, and the Brain," *Gut* 51 (2002): 885–92.

9 **of parental approval:** A vast literature exists showing how love develops and is more enduring in arranged marriages in Jewish communities and beyond. A set of two large studies conducted by Robert Epstein and colleagues on married couples from twelve different countries and six different religions has shown that love can grow over time in the absence of initial romantic involvement and has identified some of the

factors that sustain such growth. Some of these include children, commitment, thoughtfulness, forgiveness, sacrifice, and humor. See R. Epstein, M. Pandit, and M. Thakar, "How Love Emerges in Arranged Marriages: Two Cross-Cultural Studies," *Journal of Comparative Family Studies* 44 (2013): 341–60. See also "The Leap" for the relationship between time and intimacy.

10 **age for nuptials:** The data on marriage are reported from a paper entitled "Historical Marriage Trends from 1890–2010: A Focus on Race Differences," by Diana B. Elliott, Kristy Krivickas, Matthew W. Brault, and Rose M. Kreider, published on the US Census Bureau website: https://www.census.gov/hhes/socdemo/marriage/data/acs/ ElliottetalPAA2012paper.pdf. Accessed November 3, 2015.

10 **11 percent, respectively:** Here, unlike the case of median age at first marriage, the percentage of people who were never married at age thirty-five (and also at forty-five) was not lowest in the 1950s but in the 1980s. The authors of the survey argue that this might be due exactly to the early entry into marriage of the 1950s.

10 **scream of uncertainty:** For an excellent source on the desire to understand and cope with uncertainty, please refer to Helga Nowotny, *The Cunning of Uncertainty*, Polity Press, 2015.

12 **to distinguish them:** C. L. Pickett and W. L. Gardner, "The Social Monitoring System: Enhanced Sensitivity to Social Cues as an Adaptive Response to Social Exclusion," in K. D. Williams, J. P. Forgas, and W. von Hippel, eds., *The Social Outcast: Ostracism, Social Exclusion, Rejection, and Bullying*, Psychology Press, 2005, 214–26.

12 **on positive experience:** J. T. Cacioppo, J. M. Ernst, M. H. Burleson, M. K. McClintock, W. B. Malarkey, L. C. Hawkley, R. B. Kowaleski, A. Paulsen, J. A. Hobson, K. Hugdahl, D. Spiegel, and G. G. Bernston, "Lonely Traits and Concomitant Physiological Processes: The MacArthur Social Neuroscience Studies," *International Journal of*

Psychophysiology 35 (2000): 143–54. A comprehensive study that looked at differences between lonely and nonlonely adults in terms of optimism, responses to social interactions, stress coping, attention control, autonomic and neuroendocrine function, and sleep patterns.

12 **relish, social stimulation:** J. T. Cacioppo, C. J. Norris, G. Monteleone, H. Nusbaum, "In the Eye of the Beholder: Individual Differences in Perceived Social Isolation Predict Regional Brain Activation to Social Stimuli," *Journal of Cognitive Neuroscience* 21 (2009): 83–92.

13 **to overcome it:** Cacioppo et al., "Lonely Traits" (2000).

13 **a companion today:** See chapter 3 in E. Illouz, *Why Love Hurts: A Sociological Explanation*, Polity Press, 2012.

14 **made a purchase:** B. Schwartz and A. Ward, "Doing Better but Feeling Worse: The Paradox of Choice," in P. A. Linley and S. Joseph, eds., *Positive Psychology in Practice*, Wiley, 2004, 86–104.

15 **brooded about it:** As Eva Illouz illustrates in her chapter on commitment phobia in *Why Love Hurts*, these are the pitfalls of what psychologists call affective forecasting. See also T. D. Wilson, "Don't Think Twice, It's All Right," *New York Times*, December 29, 2005 (http://www.nytimes.com/2005/12/29/opinion/dont-think-twice-its-all-right.html?_r=0) and one of the original papers referred to in the article: T. D. Wilson and D. Kraft, "Why Do I Love Thee?: Effects of Repeated Introspections about a Dating Relationship," *Personality and Social Psychology Bulletin* 19 (1993): 409–18.

15 **with in reality:** A. Courtiol, S. Picq, B. Godelle, M. Raymond, and J. B. Ferdy, "From Preferred to Actual Mate Characteristics: The Case of Human Body Shape," *PLOS One* 5, no. 9 (2010): e13010.

15 **status, for instance:** D. M. Buss, "Sex Differences in Human Mate Preference: Evolutionary Hypotheses Tested in 37 Cultures," *Behavioral and Brain Sciences* 12 (1989): 1–49.

15   **the most miserable:** Schwartz and Ward, "Doing Better."

16   **experience social interaction:** H. Ruan and C. F. Wu, "Social Interaction-Mediated Lifespan Extension of Drosophila Cu/zn Superoxide Dismutase Mutants," *Proceedings of the National Academy of Sciences* 105 (2008): 7506–10.

16   **reunited with her:** As an example, see: A. Moles, B. L. Kieffer, and F. R. D'Amato, "Deficit in Attachment Behavior in Mice Lacking the Mu-Opioid Receptor Gene," *Science* 304 (2004): 1983–86. This is a study showing the importance of the opioid system in modulating the attachment behavior between a mother and her pups. Mice, as well as primates, deficient in opioid metabolism are more indifferent to their mom's absence. They experience the separation with reduced distress. Differences at the level of attachment behavior in relation to opioids are also observable in humans. See C. S. Barr, M. L. Schwandt, S. G. Lindell, J. D. Higley, et al. "Variation at the Mu-Opioid Receptor Gene (OPRM1) Influences Attachment Behavior in Infant Primates," *Proceedings of the National Academy of Sciences* 105 (2008): 5277–81; and A. Troisi, G. Frazzetto, V. Carola, G. Di Lorenzo, M. Coviello, A. Siracusano, and C. Gross, "Variation in the Mu-Opioid Receptor Gene (OPRM1) Moderates Influence of Early Maternal Care on Fearful Attachment," *Social Cognitive and Affective Neuroscience* 7 (2012): 542–47.

16   **of the pups:** C. Kuhn and S. Schanberg, "Responses to Maternal Separation: Mechanisms and Mediators," *International Journal of Developmental Neuroscience* 16 (1998): 261–70.

16   **of groundbreaking experiments:** Harry F. Harlow, "Love in Infant Monkeys," *Scientific American* 200 (June 1959).

17   **amount of food:** T. Field, M. Diego, and M. Hernandez-Reif, "Preterm Infant Massage Therapy Research: A Review," *Infant Behavior and Development* 33 (2010): 115–24.

17   **immune system:** L. M. Forsell and J. A. Åström, "Meanings of Hugging: From Greeting Behavior to Touching Implications," *Comprehensive Psychology* 1 (2012): 1–6.

17   **electric shock:** J. A. Coan, H. S. Schaefer, and R. J. Davidson, "Lending a Hand: Social Regulation of the Neural Response to Threat," *Psychological Sciences* 17 (2006): 1032–39. Activation in brain areas involved in stress response also decreased when the participants held their husbands' hands and, the happier the marriage, the better the effect.

17   **to massage infants:** T. M. Field, M. Hernandez-Reif, O. Quintino, S. Schanberg, and C. Kuhn, "Elder Retired Volunteers Benefit from Giving Massage Therapy to Infants," *Journal of Applied Gerontology* 17 (1998): 229–39.

17   **social interaction:** The neurons studied in this investigation are part of the dopaminergic system. As I explain in more detail in "The Leap," the neurotransmitter dopamine is involved in social reward and is produced in brain regions such as the ventral tegmental area when anticipating pleasure and in early phases of romantic love and attachment. In this study about the response to loneliness, the dopaminergic neurons looked at are those in the dorsal raphe nucleus. G. A. Matthews, E. H. Nieh, C. M. Vander Weele, S. A. Halbert, R. V. Pradhan, A. S. Yosafat, G. F. Glober, E. M. Izadmehr, R. E. Thomas, G. D. Lacy, C. P. Wildes, M A. Ungless, and K. M. Tye, "Dorsal Raphe Dopamine Neurons Represent the Experience of Social Isolation," *Cell* 164 (2016): 617–31.

The Leap

23   **by a second:** R. A. Nelson, D. D. McCarthy, S. Malys, J. Levine, B. Guinot, H. F. Fliegel, R. L. Beard, and T. R. Bartholomew, "The Leap Second: Its History and Possible Future," *Metrologia* 38 (2001): 509–29.

24 **"word for it":** Letters of Rainer Maria Rilke: 1892–1910, Norton, 1945.

24 **an open question:** D. M. Eagleman, P. U. Tse, D. Buonomano, P. Janssen, A. C. Nobre, and A. O. Holcombe, "Time and the Brain: How Subjective Time Relates to Neural Time," Journal of Neuroscience 25 (2005): 10369–71.

24 **of mental events:** M. I. Posner, "Timing the Brain: Mental Chronometry as a Tool in Neuroscience," PLOS Biology 3, no. 2 (2005): e51.

25 **processing of others:** I have described this in more detail in chapter 2 of my book Joy, Guilt, Anger, Love, Penguin, 2014. For a thorough revisited view on the taxonomy of mental function and dynamics of neural networks, please refer to M. L. Anderson, After Phrenology: Neural Reuse and the Interactive Brain, MIT Press, 2015. See also S. Hamann, "Mapping Discrete and Dimensional Emotions onto the Brain: Controversies and Consensus," Trends in Cognitive Sciences 16 (2012): 458–66.

25 **of its intensity:** C. E. Waugh, J. P. Hamilton, and I. H. Gotlib, "The Neural Temporal Dynamics of the Intensity of Emotional Experience," NeuroImage 49 (2010): 1699–707.

25 **several timescales:** For an overview of the temporal dynamics of mental life, brain function, and social interactions, see R. Hari and L. Parkkonen, "The Brain Timewise: How Timing Shapes and Supports Brain Function," Philosophical Transactions of the Royal Society B 370 (2015): 1–10; also, R. Hari, L. Parkkonen, and C. Nangini, "The Brain in Time: Insights from Neuromagnetic Recordings," Annals of the New York Academy of Sciences 1191 (2010): 89–109.

26 **down more slowly:** B. S. Schuyler, T. R. A. Kral, J. Jacquart, C. A. Burghy, H. Y. Weng, D. M. Perlman, et al., "Temporal Dynamics of Emotional Responding: Amygdala Recovery Predicts Emotional Traits," Social Cognitive and Affective Neuroscience 9 (2014): 176–81.

26 **a negative mood:** G. J. Siegle, S. R. Steinhauer, M. E. Thase, V. A. Stenger, and C. S. Carter, "Can't Shake That Feeling: Event-Related fMRI Assessment of Sustained Amygdala Activity in Response to Emotional Information in Depressed Individuals," *Biological Psychiatry* 51 (2002): 693–97.

26 **closeness over time:** For a review of techniques to generate closeness and intimacy, see R. Epstein, "How Science Can Help You Fall in Love," *Scientific American* (January/February 2010): 26–33.

26 **back to 1997:** A. Aron, E. Melinat, E. N. Aron, R. D. Vallone, and R. J. Bator, "The Experimental Generation of Interpersonal Closeness: A Procedure and Some Preliminary Findings," *Personal and Social Psychology Bulletin* 23 (1997): 363–77. The paper contains all the sets of questions asked during the procedure.

27 **with each other:** Research by Prof. Richard Wiseman as reported in the *Telegraph*: http://www.telegraph.co.uk/news/science/science-news/9373087/Watch-out-lotharios-Faking-romantic-feelings-can-actually-lead-to-the-real-thing.html; R. Wiseman, *Rip It Up: Forget Positive Thinking, It's Time for Positive Action*, Macmillan, 2015.

27 **at a face:** Looking into someone else's pupil engages brain areas that process reward prediction. See K. Kampe, C. D. Frith, R. J. Dolan, and U. Frith, "Reward Value of Attractiveness and Gaze," *Nature* 413 (2001): 589.

27 **a speed-dating event:** J. C. Cooper, S. Dunne, T. Furey, and J. P. O'Doherty, "Dorsomedial Prefrontal Cortex Mediates Rapid Evaluations Predicting the Outcome of Romantic Interactions," *Journal of Neuroscience* 32 (2012): 15647–56.

28 **endeavor takes effort:** Researcher Robert Epstein has extensively employed "love-building exercises" in his psychology classes, with 90 percent of students reporting improvement in their relationships

(Epstein, "How Science Can Help You Fall in Love"). Epstein has conducted studies on married couples from twelve different countries and six different religions and identified some of the factors that sustain the increase of love: children, commitment, thoughtfulness, forgiveness, sacrifice, and humor. See R. Epstein, M. Pandit, and M. Thakar, "How Love Emerges in Arranged Marriages: Two Cross-Cultural Studies," *Journal of Comparative Family Studies* 44 (2013): 341–60.

28 **favor long-term commitment:** Helen Fisher has written at length about the neurochemistry of different stages of love in her book *Why We Love: The Nature and Chemistry of Romantic Love*, Henry Holt, 2004. See also T. R. Insel and L. J. Young, "The Neurobiology of Attachment," *Nature Reviews Neuroscience* 2 (2001): 129–36.

29 **of social behavior:** I have described this in chapter 7 of *Joy, Guilt, Anger, Love*, Penguin, 2014. For a comprehensive review of the role of oxytocin and vasopressin in different kinds of affiliation behavior—from parent-child attachment to friendship and romantic bonds—see the following: R. Feldman, "Oxytocin and Social Affiliation in Humans," *Hormones and Behavior* 61 (2012): 380–91. S. Carter, "Neuroendocrine Perspectives on Social Attachment and Love," *Psychoneuroendocrinology* 23 (1998): 779–818. L. J. Young and L. M. Flanagan-Cato, "Editorial Comment: Oxytocin, Vasopressin and social behavior," *Hormones and Behavior* 61 (2012): 227–29.

29 **in their brains:** T. R. Insel, "The Challenge of Translation in Social Neuroscience," *Neuron* 65 (2010): 768–79. T. R. Insel, "Oxytocin—a Neuropeptide for Affiliation: Evidence from Behavioural, Receptor Autoradiographic, and Comparative Studies," *Psychoneuroendocrinology* 17 (1992): 3–35. According to circumstances, dopamine, oxytocin, and vasopressin increase or dampen one another's functions. For instance, oxytocin interferes with dopamine by binding to dopamine receptors in the nucleus accumbens, thereby inhibiting reward mechanisms.

29 **in defense of offspring:** With a few similar gender differences, both
men and women produce oxytocin and vasopressin; one study has
shown that mothers and fathers show the same elevated blood levels of
oxytocin during the first six months of parenting: I. Gordon, O.
Zagoory-Sharon, J. F. Leckman, and R. Feldman, "Oxytocin and the
Development of Parenting in Humans," *Biological Psychiatry* 68 (2010):
377–82. See "A Wizard's Farewell" for more on the role of oxytocin and
vasopressin in maternal and parental care, also in connection to
selective brain activity.

30 **in new parents:** I. Schneiderman, O. Zagoory-Sharon, J. F. Leckman,
and R. Feldman, "Oxytocin During the Initial Stages of Romantic
Attachment," *Psychoneuroendocrinology* 37 (2012): 1277–85. The
researchers did not measure the participants' oxytocin levels before
the formation of their bond, so they cannot ascertain whether the
elevated oxytocin was a direct result of the start of the relationship,
or a pre-existing trait that facilitated their falling in love.

30 **stages of commitment:** See L. Gravotta, "Be Mine Forever: Oxytocin
May Help Build Long-Lasting Love," *Scientific American* (February 12,
2013): http://www.scientificamerican.com/article/be-mine-forever
-oxytocin/.

30 **of stress, sank:** B. Ditzen, M. Schaer, B. Gabriel, G. Bodenmann,
U. Ehlert, and M. Heinrichs, "Intranasal Oxytocin Increases
Positive Communication and Reduces Cortisol Levels During
Couple Conflict," *Biological Psychiatry* 65 (2009): 728–31. Beyond
measurements at the hormonal level, genetic variation for oxytocin
and vasopressin validates their role in modulating individual
differences in social behavior. For a review of how variation in
oxytocin- and vasopressin-related genes modulate social behavior and
mating, see R. P. Ebstein, A. Knafo, D. Mankuta, S. H. Chew, and P. S.
Lai, "The Contributions of Oxytocin and Vasopressin Pathway Genes
to Human Behavior," *Hormones and Behavior* 61 (2012): 359–79.

31 **and elegant experiment:** P. Mangan and P. Bolinskey, "Underestimation of Time During Normal Aging: The Result of the Slowing of a Dopaminergic Regulated Internal Clock?" Paper presented at the Annual Meeting of the Society for Neuroscience (1997).

31 **in dopaminergic function:** W. H. Meck, "Neuropharmacology of Timing and Time Perception," *Cognitive Brain Research* 3 (1996): 227–42.

31 **our internal clock:** See S. Blakeslee, "Running Late: Researchers Blame Aging Brain," *New York Times*, March 3, 1998: http://www .nytimes.com/1998/03/24/science/running-late-researchers-blame -aging-brain.html?pagewanted=all&src=pm; see also N. D. Volkow, R. C. Gur, G. J. Wang, J. S. Fowler, P. J. Moberg, Y. S. Ding, R. Hitzmemann, G. Smith, and J. Logan, "Association Between Decline in Brain Dopamine Activity with Age and Cognitive and Motor Impairment in Healthy Individuals," *American Journal of Psychiatry* 155 (1998): 344–49. For evidence of general decline of dopaminergic regulation in the reward system, see J. C. Dreher, A. Meyer-Lindenberg, P. Kohn, and K. F. Berman, "Age-Regulated Changes in Midbrain Dopaminergic Regulation of the Human Reward System," *Proceedings of the National Academy of Sciences* 105 (2008): 15106–111.

31 **will go awry:** T. H. Rammsayer, "Effects of Body Core Temperature and Brain Dopamine Activity on Timing Processes in Humans," *Biological Psychology* 46 (1997): 169–92.

32 **with one another:** E. Ferrer and J. L. Helm, "Dynamical Systems Modeling of Physiological Coregulation in Dyadic Interactions," *International Journal of Psychophysiology* 88 (2013): 296–308. J. L. Helm, D. Sbarra, and E. Ferrer, "Assessing Cross-Partner Associations in Physiological Responses via Coupled Oscillator Models," *Emotion* 12 (2013): 748–62.

32 **mimicked. Gaits align:** A. Z. Zivotofsky and J. M. Hausdoff, "The Sensory Feedback Mechanisms Enabling Couples to Walk Synchronously: An Initial Investigation," *Journal of Neuroengineering*

and Rehabilitation 4 (2007): 28. Without instructions to do so, people who have never walked side by side before adopted a synchronized gait in about 50 percent of the cases in this study.

32 **synchrony catalyzes bonding:** L. K. Miles, L. K. Nind, and C. N. Macrae, "The Rhythm of Rapport: Interpersonal Synchrony and Social Perception," *Journal of Experimental Social Psychology* 45 (2009): 585–89.

32 **interaction game harmonized:** S. Cacioppo, H. Zhou, G. Monteleone, E. A. Majaka, K. A. Quinn, A. B. Ball, G. J. Norman, G. R. Semin, and J. T. Cacioppo, "You Are in Sync with Me: Neural Correlates of Interpersonal Synchrony with a Partner," *Neuroscience* 277 (2014): 842–58.

32 **more pronounced affiliation:** The players were sitting in different rooms and playing remotely via a computer. The synchronization results were backed up by brain-imaging data revealing, as one would expect, synchrony connected to the concomitant recruitment of brain regions involved in embodied cognition, action observation, as well as the processing of borders between self and others, such as the left inferior parietal lobule, the ventromedial prefrontal cortex, and parts of the parahippocampal gyrus (Cacioppo et al., "You Are in Sync," 2014).

33 **for mental simulation:** U. Hasson, Y. Nir, I. Levy, G. Fuhrmann, and R. Malach, "Intersubject Synchronization of Cortical Activity During Natural Vision," *Science* 303 (2004): 1634–40. L. Nummenmaa, E. Glerean, M. Viinikainen, I. Jääskeläinen, R. Jari, and M. Sams, "Emotions Promote Social Interaction by Synchronizing Brain Activity Across Individuals," *Proceedings of the National Academy of Sciences* 109 (2012): 9599–604. Of note, this and the next two cited studies do not consist of observations between pairs or teams of individuals watching the clips together in the same room and at the same time. Rather, they record the alignment of emotional responses across group participants who watched the clips separately. Appropriate measurement and interpretation of the synchronization are ensured by moment-to-moment recording of their brain responses to, and

subjective experiences of, the clip's emotional events during and after the screening.

33 **an interior decorator:** J. M. Lahnakoski, E. Glerean, I. P. Jääskeläinen, J. Hööna, R. Hari, M. Sams, and L. Nummenmaa, "Synchronous Brain Activity Across Individuals Underlies Shared Psychological Perspectives," *NeuroImage* 100 (2014): 316–24.

33 **between those areas:** L. Nummenmaa, H. Saarimäki, E. Glerean, A. Gotsopoulos, I. Jääskeläinen, R. Hari, and M. Sams, "Emotional Speech Synchronizes Brains Across Listeners and Engages Large-Scale Dynamic Brain Networks," *NeuroImage* 102 (2014): 498–509. The synchronicity was observed between areas in the limbic system and the prefrontal and orbitofrontal cortices.

34 **lack of warmth:** Ziyad Marar dedicates a chapter, "A Complicated Kindness," to the delicate difficult job of reaching a balance of kindness in intimate relationships in his book *Intimacy*, Routledge, 2014.

35 **well as prediction:** G. J. Stephens, L. J. Silbert, and U. Hasson, "Speaker-Listener Neural Coupling Underlies Successful Communication," *Proceedings of the National Academy of Sciences* 107 (2010): 14425–30. S. Dikker, L. J. Silbert, U. Hasson, and J. D. Zevin, "On the Same Wavelength: Predictable Language Enhances Speaker-Listener Brain-to-Brain Synchrony in Posterior Superior Temporal Gyrus," *Journal of Neuroscience* 34 (2014): 6267–72.

35 **its spoken components:** For an engaging summary of turns in conversation, see E. Young, "The Incredible Thing We Do During Conversations," *The Atlantic* (January 2016): http://www.theatlantic.com/science/archive/2016/01/the-incredible-thing-we-do-during-conversations/422439/. For a thorough scientific review, see S. Levinson, "Turn-Taking in Human Communication—Origins and

Implications for Language Processing," *Trends in Cognitive Sciences* 20 (2016): 6–14.

38 **leaned and tilted:** In a fascinating study, neuroscientist Onur Güntürkün has observed couples kiss on the lips at airports, train stations, and other public spaces in different countries. Two-thirds of them consistently tilted their heads to the right, apparently regardless, Güntürkün suggests, of being right or left handed. See O. Güntürkün, "Human Behaviour: Adult Persistence of Head-Turning Symmetry," *Nature* 421 (2003): 6924.

38 **first sexual experience:** For a comprehensive and detailed description of the anatomy, psychology, and neurochemistry behind a kiss, see S. Kirshenbaum, *The Science of Kissing*, Grand Central Publishing, 2011. Cited therein with regard to the link between kissing and memory is J. Reed, J. Bohannon, G. Gooding, and A. Stehman, "Kiss and Tell: Affect and Retellings of First Kisses and First Meetings," a paper presented at a conference of the Association for Psychological Science in Miami, Florida (2000).

Transit of Venus

40 **Transit of Venus:** I know of one novel by Shirley Hazzard, entitled *The Transit of Venus*, in which there is a reference to the astronomical event. I employ the planet's transit to elaborate on extramarital relationships.

43 **reported unfaithful behavior:** A. Kinsey, W. Pomeroy, C. Martin, and P. Gebhard, *Sexual Behavior in the Human Female*, Saunders, 1953.

43 **exceed these figures:** D. M. Buss and T. K. Shackelford, "Susceptibility to Infidelity in the First Year of Marriage," *Journal of Research in Personality* 31 (1997): 193–221.

43   **cause of divorce:** P. R. Amato and D. Previti, "People's Reasons for Divorcing: Gender, Social Class, the Life Course, and Adjustment," *Journal of Family Issues* 24 (2003): 602–26.

43   **who opposed it:** L. Appignanesi, *Trials of Passion: Crimes Committed in the Name of Love and Madness*, Virago Books, 2014.

44   **the long term:** These are long-debated and difficult questions in evolutionary studies. For a summary, see: S. M. Drigotas and W. Barta, "The Cheating Heart: Scientific Explorations of Infidelity," *Current Directions in Psychological Science* 10 (2001): 177–80. Most of the ideas reported here are those explained in an excellent book on human sexual evolution: C. Ryan and C. Jethá, *Sex at Dawn: How We Mate, Why We Stray, and What It Means for Modern Relationships*, Harper Perennial, 2010.

44   **involved no sex:** D. M. Buss, R. J. Larsen, D. Westen, and J. Semmelroth, "Sex Differences in Jealousy: Evolution, Physiology and Psychology," *Psychological Science* 3 (1992): 251–55.

44   **by sexual infidelity:** With different percentages, studies have been replicated, but there have also been conflicting reports. See M. J. Tagler and H. M. Jeffers, "Sex Differences in Attitudes Toward Partner Infidelity," *Evolutionary Psychology* 11 (2013): 821–32.

45   **her tennis instructor:** See chapter 3 in Ryan and Jethá, *Sex at Dawn*.

45   **young in communities:** Fisher, *Why We Love*.

46   **and collective paternity:** See chapter 6 in Ryan and Jethá, *Sex at Dawn*.

46   **in ancient Greece:** All these examples are reported in chapters 6 and 8 of Ryan and Jethá, *Why We Love*. Evidence about Mozambique is from a World Health Organization study conducted by one of the authors, Cacilda Jethá. The information about Iron Age Britain is from Julius Caesar's *Gallic Wars*.

50   to mating opportunities: M. McIntyre, S. W. Gangestad, P. B. Gray, J. F. Chapman, T. C. Burnham, M. T. O'Rourke, and R. Thornhill, "Romantic Involvement Often Reduces Men's Testosterone Levels—but Not Always: The Modulating Role of Extrapair Sexual Interest," *Journal of Personality and Social Psychology* 91 (2006): 642–51.

51   cheating by twofold: D. A. Weiser, D. J. Weigel, C. B. Lalasz, and W. Evans, "Family Background and Propensity to Engage in Infidelity," *Journal of Family Issues* (First published online: April 22, 2015): 1–19.

51   neuroticism and narcissism: L. Widman and J. K. McNulty, "Narcissism and Sexuality," in W. K. Campbell and J. Miller, eds., *The Handbook of Narcissism and Narcissistic Personality Disorder: Theoretical Approaches, Empirical Findings, and Treatment*, Wiley, 2011.

51   40 percent in women: B. P. Zietsch, L. Westberg, P. Santtila, and P. Jern, "Genetic Analysis of Human Extrapair Mating: Heritability, Between-Sex Correlation, and Receptor Genes for Vasopressin and Oxytocin," *Evolution and Human Behavior* 36, no. 2 (2015): 130–36. This study confirmed a previous one that looked at the variation exclusively in women and found it to be 41 percent: L. F. Cherkas, E. C. Oelsner, Y. T. Mak, A. Valdes, and T. Spector, "Genetic Influence on Female Identity and Number of Sexual Partners in Humans: A Linkage and Association Study of the Role of the Vasopressin Receptor Gene (AVPR1A)," *Twin Research* 7 (2004): 649–58.

52   on their partners: J. R. Garcia, J. McKillop, E. L. Aller, A. M. Merriwether, D. Sloan Wilson, and J. K. Lum, "Associations Between Dopamine D4 Receptor Gene Variation with Both Infidelity and Sexual Promiscuity," *PLOS* 5, no. 11 (2010): e14162.

52   of sexual behavior: A. C. Halley, M. Boretsky, D. A. Puts, and M. Shriver, "Self-Reported Sexual Behavioral Interests and Polymorphisms in the Dopamine Receptor D4 (*DRD4*) Exon III VNTR in Heterosexual Young Adults," *Archives of Sex Behavior* 45, no. 8 (2015): 2091–100.

52   **risk of divorce:** H. Walum, L. Westberg, J. M. Henningsson, et al., "Genetic Variation in the Vasopressin Receptor 1a Gene (*AVPR1A*) Associates with Pair-Bonding Behavior in Humans," *Proceedings of the National Academy of Sciences* 105, no. 37 (2008): 14153–56.

52   **nonmarried single men:** D. Scheele, N. Striepens, O. Güntürkün, et al., "Oxytocin Modulates Social Distance Between Males and Females," *Journal of Neuroscience* 32 (2012): 16074–79.

52   **have gained tolerance:** J. R. Garcia, C. Reiber, S. G. Massey, and A. M. Merriwether, "Sexual Hookup Culture: A Review," *Review of General Psychology* 16 (2012): 161–76.

53   **in recollecting them:** J. R. Escobedo and R. Adolphs, "Becoming a Better Person: Temporal Remoteness Biases Autobiographical Memories for Moral Events," *Emotion* 10 (2010): 511–18. I have described this in chapter 2 of *Joy, Guilt, Anger, Love*, Penguin, 2014.

53   **from our contradictions:** J. D. Foster and T. A. Misra, "It Did Not Mean Anything (about Me): Cognitive Dissonance Theory and the Cognitive and Affective Consequences of Romantic Infidelity," *Journal of Social and Personal Relationships* 30 (2013): 835–57.

53   **facing human nature:** J. Armstrong, *Conditions of Love*, Penguin, 2002. Please refer to the chapter on sexuality.

55   **by eight years:** To learn about the history and understand the geometrical basis of the calculations relative to the transit of Venus, consult the following educational website, which I found helpful: http://www.astronomy.ohio-state.edu/~pogge/Ast161/Unit4/venussun.html.

55   **1761 and 1767:** Better measurements were recorded in 1874 and 1882. The two most recent transits occurred in 2004 and 2012.

55   **of something else:** For a further explanation of parallax, read this blog written for the European Space Agency: http://blogs.esa.int/venustransit/2012/05/30/measuring-the-size-of-the-solar-system-parallax/.

56 **distance between them:** Triangulation was invented by the Dutch astronomer and cartographer Regnier Gemma Fresius (1508–55). See S. Christianson, *100 Diagrams That Changed the World*, Pavilion Books, 2014.

57 **"of enforced compliance":** For an exhaustive take on relationships and betrayal, see E. Perel, *Mating in Captivity: Sex, Lies and Domestic Bliss*, Hodder and Stoughton, 2007. See also "Rethinking Infidelity," Esther Perel's TED talk in Vancouver, BC, Canada (March 2015).

Split or Steal

63 **relationship poses dilemmas:** For a study that used game theory to monitor strategic decision making in heterosexual courtship, see R. M. Seymour and P. D. Sozou, "Duration of Courtship Efforts as a Costly Signal," *Journal of Theoretical Biology* 256 (2009): 1–13. For a book dedicated to the subject, see P. Szuchman and J. Anderson, *It's Not you, It's the Dishes: How to Minimize Conflict and Maximize Happiness in Your Relationship*, Random House, 2012.

64 **connection with caregivers:** J. Bowlby, *Attachment and Loss*, vols. 1–3, Basic Books, 1973. For an article on romantic love and attachment, see C. Hazan and P. Shaver, "Romantic Love Conceptualized as an Attachment Process," *Journal of Personality and Social Psychology* 52 (1987): 511–24. For an interesting and comprehensive book entirely dedicated to attachment theory in relationships, see A. Levine and R. Heller, *Attached: The New Science of Adult Attachment and How It Can Help You Find—and Keep—Love*, TarcherPerigee, 2010.

65 **from their partners:** Individuals with an anxious attachment style are exquisitely sensitive and vigilant to signs testifying to the availability and responsiveness of attachment figures. These signs may include facial emotional expressions. One study probed the ability of people who were anxious with regard to attachment to recognize the

emergence or disappearance of a given facial expression. The results indicated that, in comparison with other people, anxious individuals perceived these emotional shifts more readily. Sometimes this led to inaccuracies in emotional judgment. However, these results were reversed if anxious individuals were asked to wait before providing their answers: C. Fraley, P. M. Niedenthal, M. Marks, C. Brumbaugh, and A. Vicary, "Adult Attachment and the Perception of Emotional Expressions: Probing the Hyperactivating Strategies Underlying Anxious Attachment," *Journal of Personality and Social Psychology* 74 (2006): 1163–90 (as cited in Levine and Heller, *Attached*).

65 **wary of intimacy:** Avoidant individuals have a tendency to limit their cognitive attention to information that is related to intimacy, considering it distressful and threatening. In one study, participants were shown words on a screen and asked to name their different colored fonts. Performance at this task can be compromised by the meaning of the words—especially if negative or emotionally loaded—with participants diverting their attention from the color and focusing on the semantics. However, avoidant individuals, when presented in particular with attachment-related words such as "lonely," "intimate," "loss," "loving," or "abandon," continue to perform relatively well at the task, indicating that they skillfully ignore the meaning of the word and concentrate on the color: R. S. Edelstein and O. Gillath, "Avoiding Interference: Adult Attachment and Emotional Processing Biases," *Personality and Social Psychology Bulletin* 34 (2008): 171–81.

65 **they shun pain:** Attachment theory is one way of understanding how people deal with intimacy and behave in relationships. Furthermore, categorizing people only among three styles of attachment is restrictive. Some researchers prefer regarding these styles as positioned along a spectrum of attachment behavior; see, for instance, K. A. Brennan, C. L. Clark, and P. R. Shaver, "Self-Report Measurement of Adult Romantic Attachment: An Integrative Overview," in J. A. Simpson and W. S. Rholes, eds., *Attachment Theory and Close Relationships*, Guilford Press,

1998. Studies have found that differences in attachment styles project into a variety of traits that range from how people altruistically care for others to their emotional intelligence, level of honesty, and ability to identify and follow their true inclinations. See the following: A. Erez, M. Mikulincer, M. H. van Ijzendoorn, and P. M. Kroonenberg, "Attachment, Personality and Volunteering: Placing Volunteerism in an Attachment -Theoretical Framework," *Personality and Individual Differences* 44 (2008): 64–74. T. Lanciano, A. Curci, K. Kafetsios, L. Elia, and V. Lucia, "Attachment and Dysfunctional Rumination: The Mediating Role of Emotional Intelligence Abilities," *Personality and Individual Differences* 53 (2012): 753–58. O. Gillath, A. K. Sesko, P. R. Shaver, and D. S. Chan, "Attachment, Authenticity, and Honesty: Dispositional and Experimentally Induced Security Can Reduce Self- and Other-Deception," *Journal of Personality and Social Psychology* 98 (2010): 841–55. L. J. Otway and K. B. Carnelley, "Exploring the Associations Between Adult Attachment Security and Self-Actualization and Self-Transcendence," *Self and Identity* 12 (2013): 217–30. There also seem to be differences in sexual behavior and preferences. The avoidant might prefer concentrating on the core act or might avoid sex altogether. See chapter 10 in Levine and Heller, *Attached*, and references therein.

65 **individual's genetic makeup:** Increasing evidence points to variation in genes that code for neurotransmitters or receptors in the brain being associated to the manifestation of different attachment styles: for instance, variation in the dopamine D2 (DRD2) and serotonin 1A (5HT2A) receptors. It is crucial to bear in mind that these are only correlations with no causal value, and that no specific gene directly and univocally underlies a given attachment style. For further reading, please refer to O. Gillath, P. R. Shaver, J. M. Baek, and D. S. Chun, "Genetic Correlates of Adult Attachment Styles," *Personality and Social Psychology Bulletin* 34 (2008): 1395–405.

66 **and less anxious:** I have written about this at greater length in *Joy, Guilt, Anger, Love,* Penguin, 2014. For the papers describing the

original experiments, see the following: I. C. Weaver, N. Cervoni, F. A. Champagne, A. C. D'Alessio, S. Sharma, J. R. Seckl, S. Dymov, M. Szyf, M. J. Meaney, "Epigenetic Programming by Maternal Behavior," *Nature Neuroscience* 7 (2004): 847–54. F. A. Champagne, I. C. Weaver, J. Diorio, S. Dymov, M. Szyf, M. J. Meaney, "Maternal Care Associated with Methylation of the Estrogen Receptor-Alpha1b Promoter and Estrogen Receptor-Alpha Expression in the Medial Preoptic Area of Female Offspring," *Endocrinology* 147 (2006): 2909–15. V. Carola, G. Frazzetto, and C. Gross, "Identifying Interactions between Genes and Early Environment in the Mouse," *Genes, Brain and Behavior* 5 (2006): 189–99.

66   **hydrogen atoms, CH3:** More studies are being conducted that are aimed at unraveling several sites of methylation on the genome with consequences for individual differences for emotional regulation and social competence. One study has identified methylation for a gene encoding for an oxytocin receptor. The methylation slows transcription of the gene, resulting in lower oxytocin function and is linked to increased amygdala activity in response to negative social stimuli. M. H. Puglia, T. S. Lillard, J. P. Morris, and J. J. Connelly, "Epigenetic Modification of the Oxytocin Receptor Gene Influences the Perception of Anger and Fear in the Human Brain," *Proceedings of the National Academy of the Sciences* 112 (2015): 3308–13. Finding the molecular basis of epigenetic mechanisms is a relatively recent endeavor. Some remain cautious about discoveries in this field.

66   **and avoidant people:** For a clear and detailed explanation of the attraction between individuals with the anxious and the avoidant styles, which they call "the anxious-avoidant trap," see Levine and Heller, *Attached.*

70   **and the avoidant:** T. C. Marshall, K. Bejanyan, and N. Ferenczi, "Attachment Styles and Personal Growth Following Romantic Breakups: The Mediating Roles of Distress, Rumination, and Tendency to Rebound," *PLOS One* 9 (2013): e75161.

A Winter Garden

74   **in their circle:** For a superb essay on sex as excess, see "Sex Mad" in Adam Philips, *On Balance*, Penguin, 2010.

76   **"in the world":** D. Halperin, *Love's Irony: Six Remarks on Platonic Eros*, in S. Bartsch and T. Bartscherer, eds., *Erotikon: Essays on Eros, Ancient and Modern*. University of Chicago Press, 2005, 48–58.

76   **emotions, are subdued:** On the contrary, there is evidence that these same areas are active in the absence of erotic pleasure. For instance, individuals with hypoactive sexual desire, a condition in which desire for sexual activity is chronically low or absent, display continued activity in areas such as the medial orbitofrontal cortex when exposed to erotic images: S. Stoléru et al., "Brain Processing of Visual Sexual Stimuli in Men with Hypoactive Sexual Desire Disorder," *Psychiatry Research: Neuroimaging* 124 (2003): 67–86. See also an article by Carl Zimmer on brain and sexual desire, "Where Does Sex Live in the Brain? From Top to Bottom," *Discover* (October 2009): http://discover magazine.com/2009/oct/10-where-does-sex-live-in-brain-from-top -to-bottom.

76   **also shuts down:** B. R. Komisaruk and B. E. Whipple, "Functional MRI of the Brain During Orgasm in Women," *Annual Review of Sex Research* 16 (2005): 62–86. The lack of amygdala activity during orgasm has been studied for male ejaculation: G. Holstege, J. R. Georgiadis, A. M. J. Paans, et al., "Brain Activation During Human Male Ejaculation," *Journal of Neuroscience* 23 (2003): 9185–93.

77   **in cortical areas:** S. Ortigue and F. Bianchi-Demicheli, "The Chronoarchitecture of Human Sexual Desire: A High-Density Electrical Mapping Study," *NeuroImage* 43 (2008): 337–45.

79   **with enduring love:** For a beautiful and succinct take on the
separation of love and sex, please read chapter 20 in John Armstrong's
Conditions of Love: The Philosophy of Intimacy, Penguin, 2002.

80   **they were depressing:** Recent data suggest that millennials are having
less sex than the previous generation. See J. M. Twenge, R. A.
Sherman, and B. E. Wells, "Sexual Inactivity During Young Adulthood
Is More Common among US Millennials and iGen: Age, Period, and
Cohort Effects on Having No Sexual Partners after Age 18," *Archives of
Sexual Behavior* (2016).

81   **is not so:** In reference to the ideas about the evolutionary role of
promiscuity among heterosexuals that I described in "Transit of
Venus," I would like to add that there has certainly been a lot of
discussion about the evolutionary significance of homosexual
behavior. Not instrumental for reproduction, homosexual affiliations
have been regarded as a strongly altruistic kind of behavior, in assisting,
for instance, in caring for offspring within the same family (kin
selection). Alternatively, the homosexual trait might have persisted in
evolution because of a co-occurrence with another trait that is under
positive selection. Social and cultural pressures also play a role in
favoring homosexual ties or not. For a review of several hypotheses,
see R. C. Kirkpatrick, "The Evolution of Human Homosexual
Behavior," *Current Anthropology* 41 (2000): 385–13. For an article on
the specific case of within-family altruism tendencies and care, see an
article on Samoan communities: P. L. Vasey and D. P. VanderLaan,
"An Adaptive Cognitive Dissociation between Willingness to Help
Kin and Nonkin in Samoan Fa'afafine," *Psychological Science* 21
(2010): 292–97.

81   **their sexual boldness:** See chapter 11 in Christian Rudder, *Dataclysm:
Who We Are When We Think No One's Looking*, Fourth Estate, 2014.

81   **lifetime sex partners:** Four for gay men and straight women. Five for
lesbians and straight men. Rudder, *Dataclysm*.

82 **to sexual activity:** For a review of the advantages of social interactions and sexual activity between mates, see I. D. Neumann, "The Advantage of Social Living: Brain Neuropeptides Mediate the Beneficial Consequences of Sex and Motherhood," *Frontiers in Neuroendocrinology* 30 (2009): 483–96.

83 **lab behavioral tests:** B. Leuner, E. R. Glasper, and E. Gould, "Sexual Experience Promotes Adult Neurogenesis in the Hippocampus Despite an Initial Elevation in Stress Hormones," *PLOS One* 5, no. 7 (2010): e11597.

83 **and anxiety-reducing effects:** For a review of the involvement of neuropeptides in mediating the benefits of social interactions and sex, see Neumann, "The Advantage of Social Living," (2009).

83 **of a relationship:** E. S. Byers, "Relationship Satisfaction and Sexual Satisfaction: A Longitudinal Study of Individuals in Long-Term Relationships," *Journal of Sex Research* 43 (2005): 113–18.

83 **her sexual satisfaction:** J. H. Larson, S. M. Anderson, T. B. Holman, and B. K. Niemann, "A Longitudinal Study of the Effects of Premarital Communication, Relationship Stability, and Self-Esteem on Sexual Satisfaction in the First Year of Marriage," *Journal of Sex and Marital Therapy* 24 (1998): 193–206.

84 **in opposite directions:** D. Dentico, B. L. Cheung, et al., "Reversal of Cortical Information Flow During Visual Imagery as Compared to Visual Perception," *NeuroImage* 100 (2014): 237–43. S. M. Kosslyn, "Mental Images and the Brain," *Cognitive Neuropsychology* 22 (2005): 333–47.

85 **in deferential position:** L. Berlant and M. Warner, "Sex in Public," *Critical Inquiry* 24, no. 2, special issue on intimacy (1998): 547–66.

87 **"to satisfy it":** See Philips, *On Balance.*

88 **hear of it:** Halperin, *Love's Irony.*

88 **"sex and laughter":** Italo Calvino, "Definitions of Territories: Eroticism, Sex, and Laughter," in *The Literature Machine*, Vintage, 1997. In this essay Calvino writes primarily on how sexuality is talked about in literature but also remarks that the link between sex and laughter is important at the anthropological level.

88 **"paradoxical, and 'sacred'":** Calvino, "Definitions."

A Wizard's Farewell

95 **with a stranger:** See the following studies comparing the benefits of collaborative remembering in old-adult versus young-adult couples in recalling detailed episodic memories or performance in memory tasks: A. Rauers, M. Riediger, F. Schmiedek, and U. Lindenberger, "With a Little Help from My Spouse: Does Spousal Collaboration Compensate for the Effects of Cognitive Aging?" *Gerontology* 57 (2011): 161–66. A. J. Barniers, A. C. Priddis, J. M. Broekhuijse, C. B. Harris, R. E. Cox, and D. R. Addis, "Reaping What They Sow: Benefits of Remembering Together in Intimate Couples," *Journal of Applied Research in Memory and Cognition* 3 (2014): 261–65. C. B. Harris, A. J. Barniers, J. Sutton, and P. G. Keil, "Couples as Socially Distributed Cognitive Systems: Remembering in Everyday Social and Material Contexts," *Memory Studies* 7 (2014): 285–97.

96 **"part of motherhood":** S. Hustvedt, *Living, Thinking, Looking: Essays*, Picador, 2012.

96 **raise their offspring:** For an overview of male parental investment, please see R. Woodroffe and A. Vincent, "Mother's Little Helpers: Patterns of Male Care in Mammals," *TREE* 9 (1994): 294–97.

97 **excitement and novelty:** R. Feldman, "Infant-Mother and Infant-Father Synchrony: The Coregulation of Positive Arousal," *Infant Mental Health* 24 (2003): 1–23.

97 **neural networks:** For a general review of brain imaging and molecular studies of human parental attachment, please see J. E. Swain, P. Kim, J. Spicer, S. S. Ho, C. J. Dayton, A. Elmadih, and K. M. Abel, "Approaching the Biology of Human Parental Attachment: Brain Imaging, Oxytocin, and Coordinated Assessments of Mothers and Fathers," *Brain Research* (2014): 78–101.

97 **identify with them:** S. Atzil, T. Hendler, O. Zagoory-Sharon, Y. Winetraub, and R. Feldman, "Synchrony and Specificity in the Maternal and the Paternal Brain: Relations to Oxytocin and Vasopressin," *Journal of the American Academy of Child and Adolescent Psychiatry* 51 (2012): 798–811.

98 **of the world:** See "Love in Families" in L. Appignanesi, *All About Love*, Virago Books, 2010.

98 **with their children:** In the growing field of neuropsychoanalysis there has been interest in mapping psychoanalytic orders on the geography of the brain. It is fair to emphasize that the associations are not univocal and that, in this case, the symbolic order would not map to only one region of the brain. See chapter 2 and the references therein in M. Pizzato, *Ghosts of Theatre and Cinema in the Brain*, Palgrave Macmillan, 2006.

98 **of the brain:** For a broad description of the physiology of delirium, see J.-D. Gaudreau and P. Gagnon, "Psychotogenic Drugs and Delirium Pathogenesis: The Central Role of the Thalamus," *Medical Hypotheses* 64 (2005): 471–75. For the occurrence of delirium in connection with the use of cancer drugs, see A. Caraceni, "Drug-Associated Delirium in Cancer Patients," *European Journal of Cancer* 11, no. 2 (2013): 233–40.

100 **leads to delirium:** T. T. Hshieh, T. G. Fong, E. R. Marcantonio, and S. K. Inouye, "Cholinergic Deficiency Hypothesis in Delirium: A Synthesis of Current Evidence," *Journal of Gerontology: Medical Sciences* 63 (2008): 764–72.

101 **"of rich emotional truth":** O. Sacks, *Hallucinations*, Picador, 2012.

102 **look to care:** For a phenomenological analysis of "care" among family members who attended to relatives with dementia, see S. Peacock, W. Dubbleby, and P. Koop, "The Lived Experience of Family Caregivers Who Provided End-of-Life Care to Persons with Advanced Dementia," *Palliative and Supportive Care* 12 (2014): 117–26.

Equal

108 **and cognitive evaluation:** For an informative reading on face recognition, please see C. Maguinness and F. N. Newell, "Recognizing Others: Adaptive Changes to Person Recognition Throughout the Lifespan," in B. L. Schwartz, M. L. Howe, M. P. Toglia, and H. Otgaar, eds., *What Is Adaptive about Adaptive Memory?* Oxford University Press, 2014.

108 **the neurotransmitter dopamine:** K. K. W. Kampe, C. D. Frith, R. J. Dolan, and U. Frith, "Reward Value of Attractiveness and Gaze," *Nature* 413 (2001): 589. J. O'Doherty, J. Winston, H. Critchley, D. Perrett, D. M. Burt, and R. J. Dolan, "Beauty in the Smile: The Role of Medial Orbitofrontal Cortex in Facial Attractiveness," *Neuropsychologia* 41 (2003): 147–55.

108 **trustworthy or not:** A. Todorov, P. Mende-Siedelcki, and R. Dotsch, "Social Judgments from Faces," *Current Opinion in Neurobiology* 23 (2013): 373–80.

108 **have seen it:** J. Freeman, R. M. Stolier, Z. A. Inbretsen, and E. A. Hehman, "Amygdala Responsivity to High-Level Social Information from Unseen Faces," *Journal of Neuroscience* 34 (2014): 10573–81. Several areas of the brain are involved in these subtle evaluations, depending on the implicitness or explicitness of the judgment. Studies suggest the amygdala is particularly involved in automatic assessment of untrustworthy faces. On the contrary, when the judgment is explicit,

activity is higher in the right superior temporal sulcus, which also probes the intentionality of others. J. S. Winston, B. A. Strange, J. O. Doherty, and R. J. Dolan, "Automatic and Intentional Brain Responses during Evaluation of Trustworthiness of Faces," *Nature Neuroscience* 5 (2002): 277–83. See also Todorov et al., "Social Judgments from Faces" (2013).

109 **"distort each other":** R. Barthes, *Camera Lucida*, Vintage, 2001/1981.

109 **relation to others:** I wrote about this in my book *Joy, Guilt, Anger, Love*, Penguin, 2014. See also S. Zeki, "The Neurobiology of Love," *FEBS Letters* 581 (2007): 2575–79 (and references therein).

119 **wrote Hanif Kureishi:** H. Kureishi, *Intimacy*, Faber and Faber, 1998.

120 **Julian Barnes wrote:** J. Barnes, *Levels of Life*, Jonathan Cape, 2013.

121 **"of one another":** P. Smith, *Just Kids*, Bloomsbury, 2010.

122 **vibrate with life:** On the importance of self-fulfillment and finding one's authentic self, two interesting philosophical books are H. Frankfurt, *The Reasons of Love*, Princeton University Press, 2004; and C. Taylor, *The Ethics of Authenticity*, Harvard University Press, 1991.

122 **"attitude, follow it":** W. James, *The Principles of Psychology*, Henry Holt, 1890.

123 **with new rewards:** See C. Duhigg, *The Power of Habit: Why We Do What We Do in Life and Business*, Random House, 2012.

123 **to accept it:** Scholar Brené Brown has written on the importance of vulnerability in her book *Daring Greatly: How the Courage to Be Vulnerable Transforms the Way We Live, Love, Parent, and Lead*, Penguin, 2013.

125 **in the brain:** See the following: C. Maguinness and F. N. Newell, "Recognizing Others." P. Belin, P. E. G. Bestelmeyer, M. Latinus, and R. Watson, "Understanding Voice Perception," *British Journal of*

Psychology 102 (2011): 711–25. For a study mapping voice recognition in
the auditory cortex, see P. Belin, R. J. Zatorre, P. Lafaille, P. Ahad, and
B. Pike, "Voice-Selective Areas in Human Auditory Cortex," *Nature*
403 (2000): 309–12.

125 **groove in vinyl:** An overview, including historical references, of the
pioneering work on selective memory recall conducted in Susumu
Tonegawa's laboratory: S. Ramirez, S. Tonegawa, and L. Xu,
"Identification and Optogenetic Manipulation of Memory Engrams
in the Hippocampus," *Frontiers in Behavioral Neuroscience* 7
(2014): 1–9.

125 **across the brain:** Memory is one of the most fascinating topics in
neuroscience but also one of the most difficult to study. According to
the quality and temporal nature of the memory, different parts of
the brain are involved in storage. Without wanting to simplify too
much, it's fair to say that, for instance, the amygdala stores our fearful
memories, the hippocampus is responsible for episodic memory and for
space memory, the prefrontal cortex stores short-term memory, etc. For
a review of studies on voice memory and recognition, see D. B. Pisoni,
"Long-Term Memory in Speech Perception: Some New Findings on
Talker Variability, Speaking Rate and Perceptual Learning," *Speech
Communication* 13 (1993): 109–25.

125 **to steer behavior:** Steven Ramirez, Susumu Tonegawa, and Liu Xu at
the Massachusetts Institute of Technology have, along with their
colleagues and collaborators, produced escalating, compelling evidence
for the possibility of selectively retrieving memories by stimulating
specific cell types in the hippocampus, a region in the brain that is
involved in registering a memory's context information, as well as time
and space. The elegant set of experiments involved identifying,
implanting false memories, switching their emotional valence, and
finally even using them to ease depressive-like behavior in rodents. See
S. Ramirez, X. Liu, C. J. MacDonald, A. Moffa, J. Zhou, R. L. Redondo,

and S. Tonegawa, "Activating Positive Memory Engrams Suppresses Depression-Like Behavior," *Nature* 522 (2015): 335–39.

Yes

131 **meaning on them:** For an inspiring insight on the role of touch through the perspective of embodied-cognition and the transformative power of touch, both for the toucher and the touched, see K. Maclaren, "Touching Matters: Embodiments of Intimacy," *Emotion, Space and Society* 13, special issue on intimacy (2014): 95–102. For a detailed view on the neuroscience of touch, see also, in the same issue, J. Cole, "Intimacy: Views from Impairment and Neuroscience," 87–94.

132 **communicated to them:** M. J. Hertenstein, R. Holmes, M. McCullough, and D. Keltmer, "The Communication of Emotion via Touch," *Emotion* 9 (2009): 566–73. In this study, the accuracy rate of emotion communication through touch was between 50 and 70 percent.

132 **on average longer:** There were also differences for the type of emotion and the body location where the emotion was predominantly communicated. These were also gender-biased. For details and diagrams, see Hertenstein et al., "Emotion via Touch" (2009).

132 **of positive emotions:** L. S. Löken, J. Wessberg, I. Morrison, F. McGlone, and H. Olausson, "Coding of Pleasant Touch by Unmyelinated Afferents in Humans," *Nature Neuroscience* 12 (2009): 547–48. H. Olausson, Y. Lamarre, H. Backlund, C. Morin, B. G. Wallin, G. Starck, S. Ekholm, I. Strigo, K. Worsley, Å. B. Vallbo, and M. C. Bushnell, "Unmyelinated Tactile Afferents Signal Touch and Project to Insular Cortex," *Nature Neuroscience* 5 (2002): 900–904.

132 **"against the other":** R. Barthes, *A Lover's Discourse*, Hill and Wang, 1978.

136 **habits of intimacy:** For a review of the effects, psychological and biological, of trauma on intimacy, see B. Mills and G. Turnbull, "Broken Hearts and Mending Bodies: The Impact of Trauma on Intimacy," *Sexual and Relationship Therapy* 19 (2004): 265–89.

136 **to heal them:** For a summary on the main recent directions in the science of resilience, especially in neuroscience and epigenetics, see V. Hughes, "The Roots of Resilience," *Nature* 490 (2012): 165–67.

136 **of the brain:** K. Thomaes, E. Dorrepaal, N. Draijer, M. B. de Ruiter, A. J. van Balkom, J. H. Smith, and D. J. Veltman, "Reduced Anterior Cingulate and Orbitofrontal Volumes in Child Abuse–Related Complex PTSD," *Journal of Clinical Psychiatry* 71 (2010): 1636–44.

136 **as empathic responses:** N. I. Eisenberg, M. D. Lieberman, and K. D. Williams, "Does Rejection Hurt? An fMRI Study of Social Exclusion," *Science* 302 (2003): 290–92. C. Lavin, C. Melis, E. Mikulan, C. Gelormini, D. Huepe, and A. Ibañez, "The Anterior Cingulate Cortex: An Integrative Hub for Human Socially-Driven Interactions," *Frontiers in Neuroscience* 7 (2013), article 64: 1–4. As Hughes notes in "Roots of Resilience," the anterior cingulate cortex is also dense with opioid receptors. Natural opioids are released upon touch and are also involved in modulating attachment. See "Shidduch" and "The Leap" for more on the neurochemistry of intimacy and social interactions.

136 **is the hippocampus:** F. L. Woon, S. Sood, D. W. Hedges, "Hippocampal Volume Deficits Associated with Exposure to Psychological Trauma and Posttraumatic Stress Disorder in Adults: A Meta-analysis," *Progress in Neuro-Psychopharmacology and Biological Psychiatry* 34 (2010): 1181–88.

136 **can be extinguished:** J. Ji and S. Maren, "Hippocampal Involvement in Contextual Modulation of Fear Extinction," *Hippocampus* 17 (2007): 749–58.

137   **to the hippocampus:** N. Fani, T. Z. King, T. Jovanovic, E. M. Glover, B. Bradley, K. S. Choi, T. Ely, D. Gutman, and K. Kessler, "White Matter Integrity in Highly Traumatized Adults with and without Post-Traumatic Stress Disorder," *Neuropsychopharmacology* 37 (2012): 2740–46.

137   **family or friends:** F. Ozbay, D. C. Johnson, E. Dimoulas, C. A. Morgan III, D. Charney, and S. Southwick, "Social Support and Resilience to Stress: From Neurobiology to Clinical Practice," *Psychiatry* (May 2007): 35–40. For a study that looked at the impact of childhood abuse (of different kinds) on adult depression and how it can be modulated by social support, especially from family and friends, see A. B. Powers, K. J. Ressler, and R. G. Bradley, "The Protective Role of Friendship on the Effects of Childhood Abuse and Depression," *Depression and Anxiety* 26 (2009): 46–53.

INDEX

Joy, Guilt, Anger, Love

What Neuroscience Can—and Can't—Tell Us About How We Feel

In this engaging account, renowned neuroscientist Giovanni Frazzetto blends cutting-edge scientific research with personal stories to reveal how our brains generate our emotions. He demonstrates that investigating art, literature, and philosophy is crucial to unraveling the brain's secrets. *Joy, Guilt, Anger, Love* offers a way of thinking about science and art that will help us to more fully understand ourselves and how we feel.

"A masterful meld of science, art, and memoir on what makes us human." —Allen Frances, author of *Saving Normal*

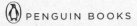